M. Inaba · Y. Inaba

Human Body Odor

Etiology, Treatment, and Related Factors

With 128 Illustrations in 222 Parts, 1 in Color

Springer Japan KK

Masumi Inaba, M.D.

Yoshikata Inaba, M.D.

Inaba Clinic (Esthetic Surgery and Dermatology) 3-31-13,
Asagaya-minami, Suginami-ku, Tokyo 166, Japan

ISBN 978-4-431-66910-4

Library of Congress Cataloging-in-Publication Data
Inaba, Masumi, 1925- Human body odor: etiology, treatment, and related
factors / M. Inaba, Y. Inaba. p.   cm. Includes bibliographical references
and index.
ISBN 978-4-431-66910-4   ISBN 978-4-431-66908-1 (eBook)
DOI 10.1007/978-4-431-66908-1
1. Bromhidrosis. 2. Hyperhidrosis. I. Inaba, Y. (Yoshikata), 1959-   ,
II. Title. [DNLM: 1. Odors–prevention & control, 2. Skin–surgery,
3. Sweat–microbiology, WV 301 I35h]. RL141,I495 1992 616.5'6–dc20
DNLM/DLC for Library of Congress   92-49858
                                                              CIP

Preface

Among the scientific studies on the sensory organs of
animal species, progress in the study of the olfactory sense
is still well behind. The fact that human beings, who con-
duct thorough studies in scientific fields, have an olfactory
sense which is duller than that of many other mammalian
species or insects seems rather strange, but it can be attributed to the fact that the human olfactory sense was
retrograded as we began to walk on our feet and the
development of the human brain preceded that of other
species. However, human beings have taken precedence in
the development of feelings of joy, anger, and personal
taste in fragrance.

Looking back over human history, we can see that a
number of races and nations experienced rises and falls in
ancient Egypt, the Orient, India, and in the vast lands of
China. The main cause of warfare was the acquisition of
precious essences which were treasured not only by the
nobles and kings as a symbol of their power but also by
ordinary people in their daily lives as well as in religious
rites. Fragrance has long been closely associated with human
life and will be as long as we possess an olfactory sense.
Among the various odors we encounter, human body odor
commonly occurs among all populations, although its relative incidence may be lower among some racial groups
than others. In a mild form, it may be treated effectively
with proper hygiene and deodorants. In an extreme form,
however, the odor may be so persistent and pervasive that
a means of permanent relief is sought. Radical treatment
procedures have been developed and employed over the
years with varied success rates.

The authors, who have developed a unique subcutaneous
shaving technique over the past 20 years, have achieved
a remarkable success record with the more than 30,000

patients treated to the present. This procedure is fully detailed here for the first time. It is hoped that the information presented will lead to new treatment breakthroughs in the future in dealing with such conditions as osmidrosis, hyperhidrosis, and bromidrosis.

Masumi Inaba
Yoshikata Inaba

Dr. MASUMI INABA is an esthetic surgeon well known in Japan and overseas for his unique surgical method which completely eliminates underarm odor. He has also propounded a new hypothesis on the subject of human hair growth and has developed a new approach to the treatment of alopecia (baldness).

Born in 1925, Dr. Inaba has practiced medicine since 1955 after graduating from Showa School of Medicine. He opened the Tokyo Research Center for Hircismus in 1965.

Dr. Inaba has written a number of published papers and books in both Japanese and English. Dr. Inaba was awarded the highest honors by the Japan Medical Association in 1979.

Dr. YOSHIKATA INABA is a dermatologist and dermatologic surgeon. He has also presented new findings on human hair growth in terms of the hair cycle by extension of his father's research. Born In 1959, Dr. Inaba has practiced medicine since 1984 after graduating from The Jikei University School of Medicine. He is now a research associate in the Jikei University department of dermatology under Professor M. Niimura.

Contents

Human Body Odor

Chapter 1. Introduction

Human body odor, and axillary odor in particular, can be a factor in human communication. Natural human odors can also provide a significant source of information involving body processes. Common sources of human body odor include apocrine and eccrine bromidrosis, halitosis, and vaginal odor, among others. The most common of these conditions is axillary apocrine bromidrosis. This condition per se is characterized by excessive perspiration and excessive, usually offensive, odor emanating from the skin by the action of surface bacteria.

The skin odor results from a breakdown in secretions mainly from the sebaceous, apocrine, and eccrine glands (Fig. 1.1). Bacterial lipases hydrolyze triglycerides to glycerol and fatty acids which are further metabolized to odoriferous compounds. Excessive sweating unaccompanied by a "goaty odor" is a condition of hyperhidrosis. Excessive odor alone, originating in the apocrine glands, is a condition of osmidrosis. The combination of both factors is defined as bromidrosis. Eccrine bromidrosis stems from bacterial action on moist keratin: for example, on the palms of the hands or the soles of the feet. If limited to excessive sweating with no detectable odor, the condition is called volar hyperhidrosis. Eccrine bromidrosis in the feet is found in all races and age groups and is about equal in both sexes if local hygiene is generally the same. It is worse in warm climates or seasons and in conditions conducive to heavier eccrine sweating.

Halitosis is a condition of malodor produced in the mouth and lungs. It may not indicate a serious medical problem, but the odor is repugnant to most people. It may be so distressing that it triggers neurosis or causes compulsive behavior typified by incessant use of mouthwashes, peppermint drops, and oral spray deodorants in order to conceal the odor and continue to win social acceptance.

Genital odor consists of genital odor per se as well as the odor of vaginal discharge in females. Genital bromidrosis, however, is less frequent than its axillary counterpart. Foul odor in vaginal secretion comes from certain complications such as tumors or inflammations that alter the normal composition of vaginal discharge.

In the diagnosis of diseases, the most important reference to body odors concerns diseases of infants that involve odors in amino acid metabolism. Intense and uncommon odors are detected in infants' breath, sweat, and urine. Systemic disease processes such as gastrointestinal disorders and diabetic ketoacidosis (acetone breath) can be revealed in odors associated with breath and/or saliva. Uremic breath has a "fishy" or "ammoniacal" odor.

The most noticeable body odor of all is that of axillary apocrine bromidrosis. Bromidrosis patients, however, show concern not only about the foul odor, but also about the excessive sweating that frequently results in clothing discoloration.

The authors have established a close correlation between axillary bromidrosis and wet cerumen in Japanese cases. If the cerumen is dry, bromidrosis almost never occurs. The basic phenomenon of cerumen moisture follows Mendelian dominant inheritance. Wet cerumen in the external auditory meatus is due to the local presence of the so-called ceruminous apocrine glands. On the other hand, axillary bromidrosis is due to the condition of the apocrine sweat gland affected by hormonal influences. Full development of the apocrine glands in the axilla does not begin until puberty (under hormonal influence), but their development in the ear begins immediately after birth.

A condition of axillary hyperhidrosis can be classified into two subtypes, essential and symptomatic. If the cerumen is dry, essential hyperhidrosis has almost no odor and is only a condition of excessive sweating that does not stain clothing. Symptomatic hyperhidrosis is always accompanied by the "goaty odor" of apocrine bromidrosis and clothing discoloration.

Apocrine bromidrosis is less common in Oriental populations than in other parts of the world. In areas where the condition is far more common, it may be "treated" on a daily basis with careful personal hygiene and topical application of deodorant and antiperspirant preparations. Since the rate of bromidrosis in Japan is only approximately 10% of the whole population, those Japanese in which it does occur become extremely concerned about it. They consider it "abnormal" and, as such, socially reprehensible. It is a

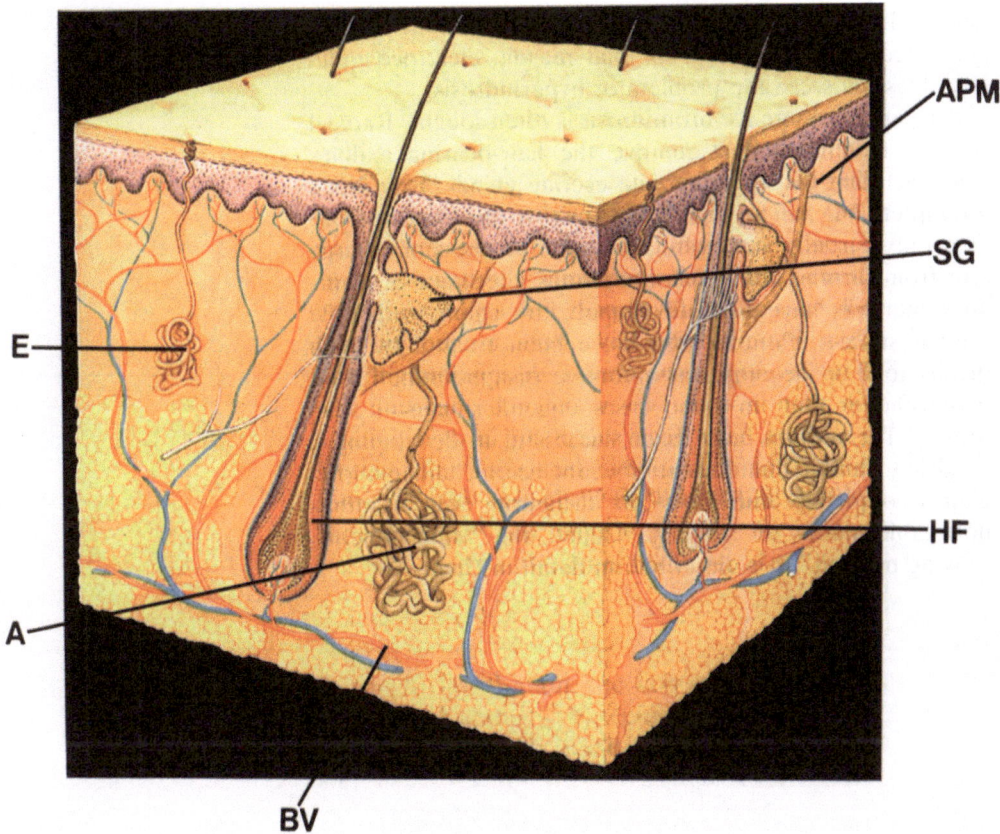

Fig. 1.1. Schematic representation of human skin. *E*, eccrine sweat gland; *A*, apocrine sweat gland; *BV*, blood vessels; *HF*, hair follicle; *SG*, sebaceous gland; *APM*, arrector pili muscle. (Modified with permission from Inaba 1984)

fact that in the public service sector, Japanese applicants afflicted with bromidrosis have been denied the chance to sit for examinations to enter the National Police Academy and to find employment as municipal firemen.

Several kinds of approaches have been developed for the treatment of osmidrosis, hyperhidrosis, and bromidrosis. The aforementioned anti-perspirants and deodorants can be effective for temporarily diminishing the odor and sweat flow, but this form of treatment is only useful for mild bromidrosis.

Electrocoagulation has been used to treat axillary bromidrosis by epilation of axillary hair. Destruction of the hair follicles is intended to diminish the apocrine sweat glands appended to them. However, the ducts of eccrine sweat glands all open individually on the skin surface, and are not appended to hair follicles; for that reason, electrocoagulation does not solve the problem of hyperhidrosis.

A permanent cure for bromidrosis is often sought. Radical surgical techniques that remove the hair-bearing axillary skin, including the apocrine and eccrine glands, have been attempted, but fail to completely remove the skin and are thus only partially successful. In order to prevent a large scar from forming after such operations, curettage or clearance methods have been developed, but again, with only limited success. Patients who have been treated by such means tend to develop postoperative disappointment and dissatisfaction that in some cases engender neurotic disorders. The authors have been successful in developing a surgical technique that does remove the eccrine and apocrine glands completely and leaves no disfiguring scar or other noticeable marks of surgery. Called the "subcutaneous shaving method," it is described in detail in Chap. 15.

Chapter 2.   Skin Structure and Bromidrosis-Related Physiology

In this chapter we will examine physiological features and functions related to bromidrosis, beginning with the subject of skin structure.

2.1   Skin Structure

In rough classification, the skin consists of two layers: epidermis and dermis. The epidermis is the surface layer exposed to the outside world. The dermal layer lies beneath it and above the subcutaneous tissue level. Epidermal thickness over the body is ordinarily 0.1–0.15 mm, and is thus quite thin, except for the heel of the foot and the palm of the hand (heel epidermis being about 1.3 mm and that of the palm about 0.7 mm). The dermis, in general, is thicker than the epidermis, measuring 1–2 mm (Fig. 2.1).

The epidermis itself has several sublayers. These include the basal layer, above which lies the dermal layer, the spinous and granular layer, and the cornified layer which is on the actual skin surface. A clear layer is observed between the granular and cornified layers in the epidermis of the palm and heel (Fig. 2.2).

Mitotic activity (cell division) occurs in the basal layer. The divided cells migrate upward to the spinous or granular layers while growing and are finally keratinized to form the cornified layer. In due time they expire and are discarded from the skin surface. Their "life cycle" from cell division to final discard is about 4 weeks.

The skin contains three kinds of glands: eccrine, apocrine, and sebaceous. All three have a relationship with bromidrosis. Schiefferdecker (1922) classified the sweat glands of the human skin into two groups, the apocrine and eccrine sweat glands, based on their mechanisms of secre-

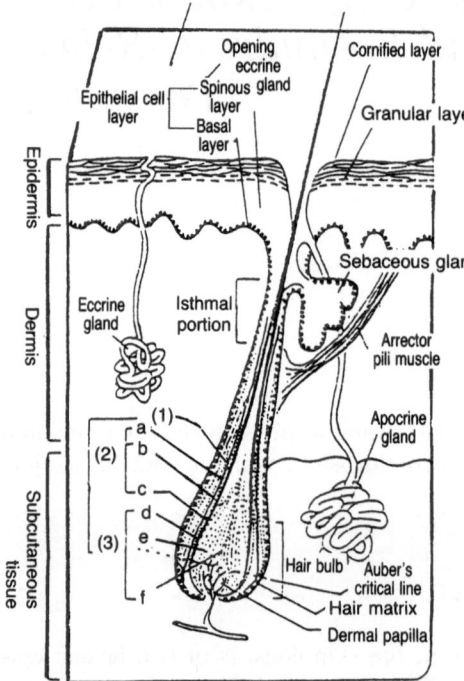

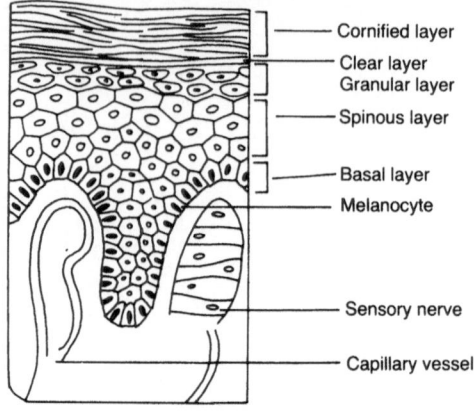

Fig. 2.2. Cross section of the epidermis. (Reproduced with permission from Inaba 1985)

Fig. 2.1. Skin and hair follicle structure. (1) Outer root sheath; (2) Inner root sheath: *a*, Henle's layer; *b*, Huxley's layer; *c*, sheath cuticula; (3) Hair: *d*, hair cuticula; *e*, cortex; *f*, medulla. (Reproduced with permission from Inaba 1985)

tion release. The eccrine glands open independently and directly onto the skin surface, whereas apocrine glands are appended to hair follicles. The duct opens slightly superior to the sebaceous duct. The sebaceous glands are associated with hair follicles and empty through a short duct into the follicular canal (Fig. 2.1).

Apocrine sweat is sterile and odorless when it first appears on the skin surface. Bacteria degrade the sweat so that the classic acrid odor is detectable. Sebum secreted from the sebaceous gland and eccrine sweat also contribute to the acrid odor.

2.2 Apocrine Gland Units

Most of the apocrine sweat glands are involved in odorous secretion functions which are of importance to the reproductive and social lives of animals. The fact that development of some apocrine glands is dependent on sexual hormones makes it evident that their functions are significant in reproductive behavior. The odors of some apocrine glands may act as pheromones.

The apocrine glands in the human body are found in the axilla, areolae, periumbilical perineal, circumanal areas, prepuce, scrotum, mons pubis, labia minora, external meatus ceruminous glands, and eyelids (Moll's glands). Apocrine glands are also distributed on the general body surface of mammals. Apocrine glands are larger and more numerous in the axilla than anywhere else; indeed, except for the axilla, the others are mere vestiges in humans.

Wet cerumen in the external meatus is a product of apocrine glands (ceruminous glands). The primordial apocrine glands, which develop after 5 months of intrauterine life, precede the development of the eccrine glands. The apocrine glands are less active before puberty than afterward. The glands are enlarged by hormonal influence and diminish much later in advanced age. This occurs in females at approximately 45 years of age and in males older than 60 years, but the glands do not disappear completely.

Ceruminous glands are developed at birth, whereas apocrine glands in other regions are affected by hormones. The anlagen of both ceruminous and axillary glands appear earlier in fetuses, but ceruminous gland function begins shortly after birth as opposed to the secretion of axillary glands, which is active in puberty due to hormonal influence. It is questionable whether axillary odor is controlled by a third allele at the same locus or whether a modifier gene or genes at another locus may be involved (Matsunaga et al. 1954).

2.2.1 Histological Structure

The apocrine unit consists of a coiled secretory portion (glomerulum) located in the lower dermis or subcutaneous fat and a straight excretory duct that empties into the hair follicle infundibulum just above the entrance of the sebaceous duct.

In cross section, the apocrine secretory coil shows a diameter about tenfold that of the eccrine secretory coil. The lumen of the apocrine secretory gland has a single cell-layer lining. The cells are varied in shape from cuboidal to columnar. Pale-staining cytoplasm is abundant, and their round nuclei are situated near the cell bases. The secretory cells have convex apical borders that project into the lumen

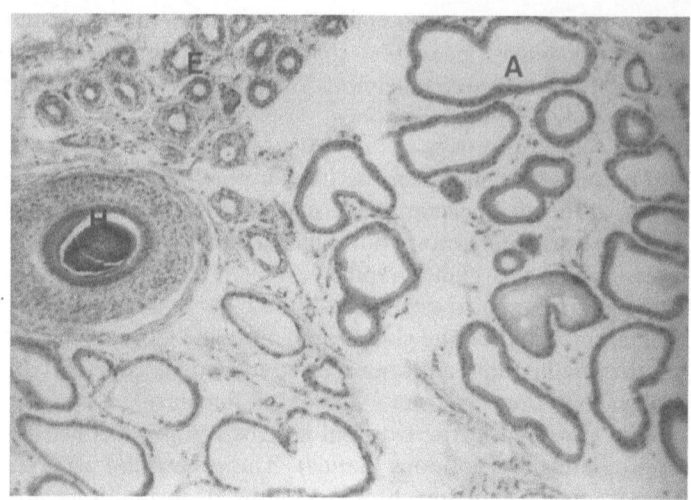

Fig. 2.3. Histologic finding in the axilla. The apocrine secretory coil shows a diameter about tenfold that of the eccrine secretory coil. *H*, hair follicle; *E*, eccrine gland; *A*, apocrine gland

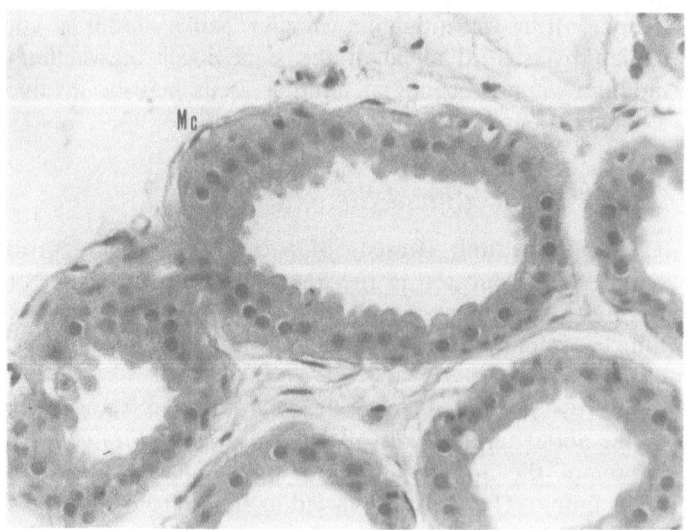

Fig. 2.4. The lumen of the apocrine secretory gland has a single cell layer lining

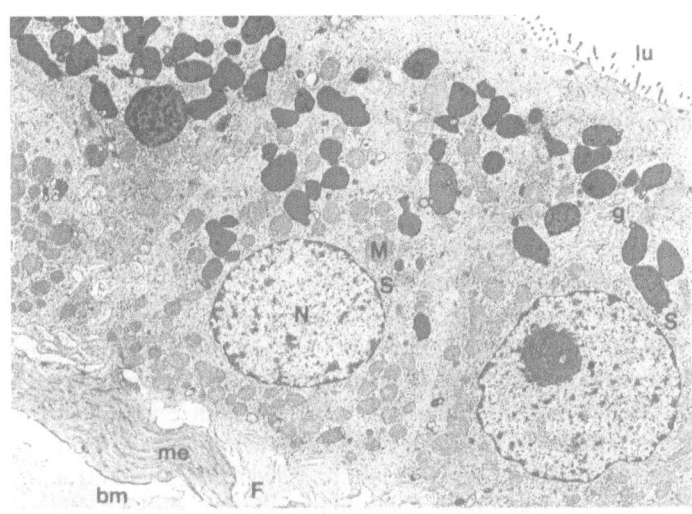

Fig. 2.5. Electron micrograph of an apocrine sweat gland from the axilla of an adult human. Two types of cells are present: secretory cells (*s*) and myoepithelial cells (*me*). The nucleus (*N*) of the secretory cell is situated at a position slightly basal from the center of the columnar secretory cells. In the supranuclear region, a number of enormous mitochondria (*M*) and large dense granules (*g*) are observed. The luminal cell surface (*lu*) is provided with microvilli. The myoepithelial cells surround the secretory cells. Gaps between the myoepithelial cells are filled with projections of secretory cells with ample foldings of cell membranes (*F*). *bm,* basement membrane; *D, dermis,* ×3900

to degrees that vary by the particular stage in the cells' secretory cycle (Figs. 2.3, 2.4). A layer of contractile myoepithelial cells, a PAS-positive basement membrane zone, and type 3 collagen and elastic fibers of the periadnexal dermis surround the secretory cells.

The apocrine secretory duct consists of a double layer of cuboidal cells that have an inner periluminal cuticle but no outer myoepithelial lining. Distally, the apocrine duct epithelium merges with the epithelium of the hair follicle's infundibular portion and cornifies independently.

2.2.2   Ultrastructural Features

Ultrastructural features of apocrine secretory cells are typical of secretory epithelium (Fig. 2.5). These features include protuberant endoplasmic reticulum and Golgi apparatus in addition to numerous ribosomes, mitochondria,

granules, and lysosomes. Large numbers of secretory granules are observed close to the luminal cell border and usually
seem to be "pinched off" into the lumen along with a small
cytoplasmic rim. The mechanism by which secretion is released from apocrine gland cells has been ascribed mainly to
apocrine secretion, also called decapitation. Small secretory
granules or vesicles may be discarded into the lumen together with the cytoplasmic matrix by the apocrine secretion
mechanism (Kurosumi et al. 1982).

The lowermost parts of secretory cells almost connect
with the myoepithelial cells, but the partial secretory cells,
especially in the human axillary apocrine gland, show a
terminal basal infolding (villous folds) that extends to the
basal lamina between the myoepithelial cells (Kurosumi et
al. 1959). However, this cannot be observed in the ceruminous apocrine gland, most likely because the ceruminous
glands secrete sweat that has little water content as distinct
from the sweat secretion of the axillary glands. Except for
this difference, no microscopic distinction appears to exist
between axillary apocrine glands and ceruminous apocrine
glands (Kawabata 1964).

The peripheral cells in a portion of the apocrine duct
located close to the hair follicle contain so many lipid droplets that they resemble sebaceous gland cells (Kurosumi
1977). Kurosumi reported that both the apocrine sweat
glands and sebaceous glands empty their secretions into the
hair follicles. Therefore, these two glands may be inter-
related embryologically, with the accumulation of numerous
lipid droplets in the apocrine duct conceivable as a kind of
metaplasia approaching the sebaceous glands.

The animal-fat lipid level in the apocrine duct and consequent rate of bromidrosis are generally much higher in
Caucasian than in Oriental populations, and this may be
due to a real difference in dietary habits, i.e., the amount of
meat consumed.

2.2.3   Mechanism of Secretion

Apocrine gland secretion is continuously formed and
periodically excreted by contraction of the secretory coil's
myoepithelial lining. Myoepithelial contraction and apocrine secretion can be induced by oxytocin and by emotional stress that promote sympathetic nervous discharge.
The apocrine sweat glands of humans respond to emotive
stimuli only after puberty and can be more readily stimulated by either epinephrine or norepinephrine (catechol-
amine) given locally or systemically than by acetylcholine.

Studies on other species have shown that the apocrine glands are controlled by adrenergic nerves, which is in marked contrast to the eccrine glands which are under cholinergic control. Cannulation of the duct of the human apocrine sweat gland has shown that secretion is pulsatile, and it is assumed that contractions of the myoepithelial cells which surround the secretory cells are responsible for these pulsations.

2.2.4 Apocrine Sweat

Apocrine secretion is odorless and sterile, with a pH of 5.0. The apocrine sweat rapidly dries, and eccrine sweat assists in its evaporation. The apocrine gland is activated in the third trimester and begins to secrete a milky fluid comprised of water, lipids, reducing sugars, Fe^{2+}, and NH_3 (Solomons 1962).

Apocrine sweat components include proteins, pyrodextrose, iron ion, ammonia, and the lipofuscin pigments responsible for clothing stains (Hurley and Shelley 1960). Lipofuscin pigments in chemical definition are a heterogenous product of malonaldohyde, peroxidized lipids, and proteins polymerized to various degrees. These pigments are formed by the enzymatic process of lipid peroxidation and are auto-fluorescent, extremely insoluble, and inert (Siakotas et al. 1972; Williams and Howard 1983).

The apocrine glands are less active before puberty than afterward. In females, apocrine gland enlargement by hormonal influence occurs after breast enlargement and earlier than in males. Enlargement of female apocrine glands is further influenced by the menstruation cycle and pregnancy.

Harada and Inaba (1985) have reported a significant difference between the apocrine glands of bromidrosis patients and those of nonbromidrosis controls (see Chap. 6.3).

2.3 Eccrine Gland Units

The eccrine gland is the essential sweat gland. Eccrine sweat is a hypotonic solution and helps to cool the body by evaporation. Eccrine sweat units are found mainly in horses and the higher primates. They are found everywhere on human skin except at the mucocutaneous junctions (lips, external auditory meatus, prepuce, glans, labia minora, and clitoris) and are concentrated most densely on the palms, soles, axillae, and forehead.

The eccrine gland, derived from the epidermis, is situated at the juncture between the dermis and the subcutaneous fat layer, or less often, in the lower third of the dermis. The total number of eccrine sweat glands varies according to different reports. The human infant at birth usually has approximately 3 million eccrine sweat units, and no more are formed afterward. In the adult, the number is greatest on the soles of the feet and least on the back. The size of the eccrine gland varies regionally and individually, ranging between 30 and 40 µg weight in humans as well as in monkeys.

Each unit consists of a hollow tube that is distally delimited by its aperture on the skin surface and proximally by a cul-de-sac. Besides the excretion of a clear, hypotonic, and watery solution, eccrine secretions are known to contain sodium chloride, vitamins, glucose, acid, urea, potassium, fat molecules, and fatty acids (Ellis 1975). Human eccrine sweat has a hypotonic characteristic that may be active duct cell absorption of sodium and chloride from precursor sweat (Schwartz and Thaysen 1956).

2.3.1 Histological Structure

The eccrine unit can be said to consist of a secretory gland that is irregularly coiled in its proximal portion, a dermal duct that leads from the secretory glands, and a spiraled intraepidermal duct (acrosyringium) that opens on the skin surface.

The coiled secretory gland has two layers of cells: a discontinous outer row of cylindrical-shaped contractile myoepithelial cells and an inner row of pyramidal-shaped secretory epithelial cells (Figs. 2.6, 2.7). Ito (1943) recognized the presence of two different types of secretory cells in the secretory coil of the human eccrine sweat gland and named them "superficial and basal" cells, depending upon their disposition in the secretory epithelium. Montagna and his associates (1953) called those two types of secretory cells of eccrine sweat glands "dark" and "clear" cells based upon their stainability in the preparations for light microscopy.

Munger (1971) became skeptical of the theory proposed by Schiefferdecker (1917) stating that the sweat glands should be classified into two groups, the apocrine and eccrine glands. He asserted that the difference between the two glands lies in the presence or absence of clear cells, i.e., the apocrine gland can be defined as being the eccrine gland devoid of clear cells.

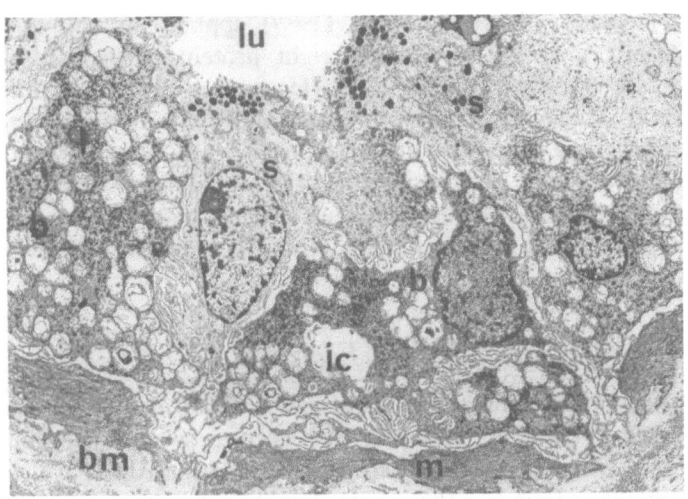

Fig. 2.6. Electron micrograph of a portion of the secretory coil of an eccrine sweat gland. Muciginous "superficial" cells (*s*) border the lumen (*lu*) while "basal" serous cells (*b*) are more deeply situated and surround intercellular canaliculi (*ic*). The basal cells have many glycogen dense secretory granules. The myoepithelial cells (*m*) rest on the basement membrane (*bm*). *l*, lipid inclusion, ×3900

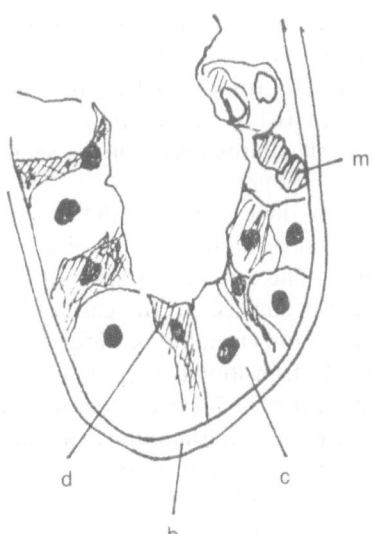

Fig. 2.7. The eccrine sweat gland has a histological structure quite similar to the apocrine gland, the main difference being the presence of a smaller glandular duct and two types of cell: the dark (superficial) cell (*d*) and the clear (basal) cell (*c*). *b*, basal membrane; *m*, myoepithelial cell

Recently, Komatsu et al. (1989) applied an antibody against a 15,000 molecular-weight protein purified from breast cyst fluid (GCDFP-15) (Haagensen 1979) to cutaneous tissue, especially to sweat glands. The normal apocrine gland cells were strongly positive and eccrine gland dark cells were also positive. Eccrine gland clear cells were weakly positive or negative. Duct cells of both glands were negative. The above finding tends to support Munger's conclusion.

2.3.2 Ultrastructural Features

The entire eccrine apparatus consists of four main portions: (1) the spiraling intraepidermal eccrine sweat duct unit (acrosyringium), (2) the straight intradermal duct, which actually continues the spiral but has an unpredictable course, (3) the coiled duct, and (4) the secretory portion. The latter two form nearly equal parts of the sweat coil, so that the two old terms, "syrinx" for the duct and "spirema" for the coil, overlap in meaning. In skin containing relatively numerous eccrine glands, the coils are found at two levels: superficial ones in the deep dermis and deep ones in the subcutaneous tissue. Pinkus and Mehregan (1981) and Kurosumi et al. (1982) reported that the entire duct system of the eccrine gland could be divided into four parts: a transitional portion, a coiled duct, a straight duct, and an epidermal duct (Fig. 2.8). The transitional portion is a short portion inserted between the secretory portion and the coiled duct, and contains structures between the secretory coil and the coiled duct.

The coiled duct and straight duct make up the dermal duct. The coiled part of the dermal duct is involved in forming secretory glomerulum. The rest of it is straight and oriented upward toward the skin surface. In the case of the eccrine gland, the duct does not usually participate in glomerulum formation, and its transitional section is conspicuous. The coiled duct has a double layer of luminal and peripheral epithelial cells, but no myoepithelium (Kurosumi et al. 1982). We suggest that regeneration of sweat glands after removal from beneath the axillary skin is based on whether the coiled duct remains. The eccrine dermal duct has a double-layer lining of small, cuboidal, and darkly basophilic epithelial cells. The luminal margin of the entire duct is rimmed by a homogeneous eosinophilic cuticle. Each of these ducts penetrates the epidermis at the nadir of a netlike ridge, then coils and expands in a spiral form up to an opening on the skin surface.

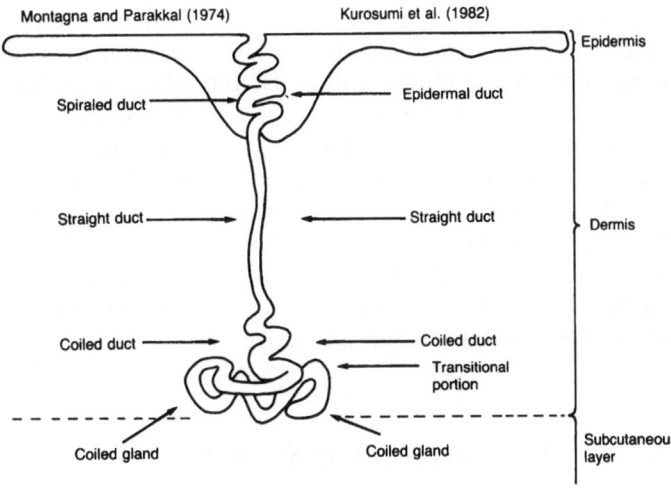

Fig. 2.8. In comparison with Montagna and Parakkal's classification, Kurosumi et al. show that the entire duct of the eccrine gland can be divided into four parts: a transitional portion, a coiled duct, a straight duct, and an epidermal duct

2.3.3 Mechanism of Secretion

Perspiration begins in the pale cells of the eccrine gland's secretory portion. Unmyelinated cholinergic nerve fibers of the sympathetic nervous system innervate the eccrine secretory coil's secretory and myoepithelial cells. A small number of unmyelinated adrenergic nerve terminals can be observed near the gland.

The eccrine glands on the palms, soles, axillae, and forehead differ from the eccrine glands elsewhere on the body surface by responding mostly to emotional as opposed to thermal stimuli.

In cases of anhidrotic ectodermal dysplasia, the patients cannot sweat. Those who live in warm climates suffer from severe heat prostration and even heat stroke, unless kept cool by other means, such as air conditioning.

Eccrine glands produce sweat in response to chemical stimuli that include acetylcholine, calcium, prostaglandin E_1, and epinephrine. The glands are rapidly responsive to stress caused by heat. In the Japanese language, psychosomatic sweating is called "shaking-hand sweating," induced, for instance, by a school examination or other stress-producing situations. Because of this stress, apocrine and eccrine sweat increases, gives rise to body odor, and influences blood circulation.

Eccrine and apocrine sweat glands of patients afflicted with axillary hyperhidrosis are no different from those of individuals without that condition. No differences have been found in staining for Schiff's-reactive material, glycogen, lipids, and metachromasia, or in visualization by fluorescent microscopy and electron microscopy. Furthermore, no variation has been detected in the innervation of these glands. Acetylcholinesterase, both in amount and localization, is the same as that in individuals whose axillary sweat glands can be considered normally responsive. Serum cholinesterase levels also lie within the normal range.

2.3.4   Demonstration of Sweating

Sweating patterns have been outlined with a simple technique, using iodine and starch, which provides a topographic, semiquantitative determination of the sweating responses colorimetrically. In this method, a 3% iodine and 3% potassium iodide solution in 95% ethyl alcohol is painted on the shaved axillae and allowed to dry. Cornstarch powder is then dusted over the painted area. As the sweat droplets appear at their ductal orifices, they make the iodine soluble and react with the cornstarch, producing a blue-black color at each pore. In areas of active sweating, the droplets become confluent, producing solid zones of blue-black color.

2.4   Sebaceous Gland Units

Sebaceous glands are found everywhere on human skin except for the palms of the hands and soles and dorsa of the feet. Mostly associated with hair follicles, these glands are emptied through a short duct into the follicular canal. With regard to mucous membranes, much variation is found in the numbers and sizes of the sebaceous acini attached to the pilosebaceous unit. The size and density of the glands are greatest in the face and scalp, ranging up to 400–800 glands/ cm^2. Gland density is usually less than 50 glands/cm^2 on the extremities (Yamada 1932).

The glands are well developed in the human embryo. Sebum, the product of sebaceous glands, is a significant contributor to the vernix caseosa. Sebaceous glands become involuted after birth and remain small in prepubertal life. The glands are later enlarged with the approach of puberty. All such changes are hormone-controlled.

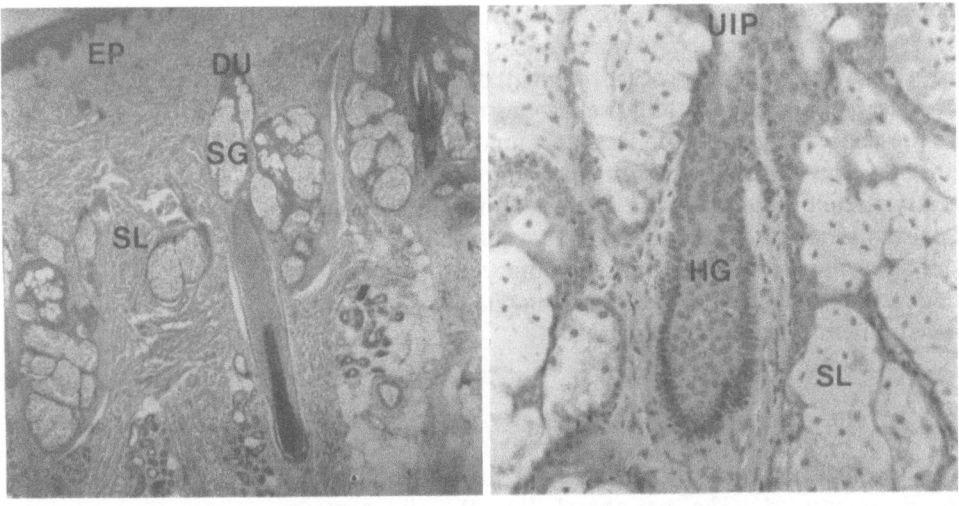

Fig. 2.9.a,b. Sebaceous gland structure. The cells surrounding the hair follicle are analogous to the cells found in the basal epidermal level. *EP*, epidermis; *HF*, hair follicle; *HG*, new hair germ; *SL*, sebaceous lobule; *UIP*, upper isthmal portion; *DU*, secretory duct; *SG*, sebaceous gland

2.4.1 Sebaceous Gland Structure

Downing et al. (1987) describe sebaceous glands as epithelial cell lobules that differentiate in the direction of lipid-producing cells in a centripetal fashion. Within the sebaceous lobule is an outer layer of undifferentiated, rather flattened germinative cells that have a great deal of homogeneous basophilic cytoplasm.

These cells are analogous to the cells found in the basal epidermal layer (Fig. 2.9.a,b). Lipids accumulate when these cells differentiate and at length fill the cell bodies. Enlarged cells observed in the central sector of the sebaceous lobule have an accordingly characteristic foamy pale-staining cytoplasm and scalloped nuclei due to compression by lipid vacuoles. When displaced toward the lobule center, the vacuolated cells slowly disintegrate into a formless mass of lipid and cellular debris (sebum) eventually discharged into the sebaceous duct.

Cornifying squamous epithelium lines the sebaceous duct. Its granular layer gradually disappears as the duct wall thins and changes with evident differentiation to lipid-producing sebaceous cells.

Most nerves do not innervate human sebaceous glands. As their cells mature, vacuoles full of lipid continue to enlarge and may even fuse with one another. When and as the cell membranes become disorganized, the cells erupt and discharge lipid and other remnant debris of opaque cytoplasmic organelles into the sebaceous duct.

The sebaceous gland is defined as a holocrine gland, due to its specific process of sebum secretion. All of the sebaceous cell and its contents are discarded into the sebaceous duct similar to the process in which desquamating cornified cells are produced by epidermal keratinocytes. "Sebum," as the secretory product of the sebaceous gland, is similar to the cornified layer of the epidermis.

2.4.2   Ultrastructural Features

Seen through the electron microscope, three types of cells are demonstrated. The peripheral (basal) cells contain abundant smooth and rough endoplasmic reticulum (ER), free ribosomes, glycogen particles, mitochondria, and from 60- to 80-A° filaments. A Golgi zone is present, but there are very few lipid droplets in the basal cells. The partially differentiated cell may have a highly convoluted plasma membrane. Its cytoplasm contains more smooth endoplasmic reticula, and there are membrane-limited lipid droplets of various sizes. It is believed that both the Golgi apparatus and the smooth endoplasmic reticulum contribute to the formation of lipid droplets.

2.4.3   Sebaceous Gland Development

The development of the glands is observed in the 13–15th week of gestation from a bulge on the follicular primordium. Then the central cells of the bulge become lipid-producing and disintegrate, leaving a lumen. The cells around the neck of the bulge keratinize to form the sebaceous duct. Cells at the apex of the bulge continue to divide and produce the sebaceous acinus containing differentiating cells. High activity is observed during the third trimester and for some months postnatally. This activity is stimulated by androgens, the production of which is quite high perinatally but declines by 1 year of age, remaining low until the approach of puberty.

As stated, sebaceous glands are rather well developed at birth but then begin to regress. At about 8–10 years of age, the progressive enlargement and accelerated productivity of these glands indicate the pre-preparation stage for puberty. This activity begins to abate in females after menopause and in males after their seventies. Sebaceous gland development and stimulation of sebogenesis are both hormone-dependent activities. Adrenal androgens in adult females may participate in the maintenance of sebaceous gland function. The ovaries play a role in continuous sebum production. Testosterone, androstenedione, and dehydroepiandrosterone secreted by the adrenal gland, are the major sebotrophic stimuli in males and do not respond to acetylcholine or norepinephrine. Sebaceous gland secretion continues to the skin surface under the control of androgens.

2.4.4   Components of Sebum

The major components of sebum are triglycerides, wax ester, squalene, cholesterol esters, and cholesterol, respectively. Only the squalene and wax ester lipids are unique to the sebaceous glands.

Kellum (1967) reported triglycerides, wax esters, and squalene, but no cholesterol or cholesterol esters in isolated sebaceous glands. In subsequent studies, however, small amounts of cholesterol and cholesterol esters have been detected in the isolated gland (Stewart 1982).

The free fatty acids are produced by the breakdown of triglicerides by lipases secreted by bacteria that normally reside in the follicular infundibula. The rate of bacterial hydrolysis determines the considerable variation observed in percentages of free fatty acids and triglycerides. Freshly secreted sebum is presumed to contain only esterified fatty acids. Esterase activity occurring in the follicular canal and on the skin surface is believed to release the free fatty acids.

2.4.5   Human Surface Sebum

One of the complications that arise in any attempt to characterize sebum is that pure glandular material is not easily obtained. Most analyses have depended on the skin surface lipid. However, this lipid is a mixture of sebaceous and epidermal lipids, even though most of the lipids are sebaceous in origin. Analyses of the surface lipid have disclosed the presence of free fatty acids, triglycerides, diglycerides, monoglycerides, wax esters, sterol ester, sterols, and squalene (Downing et al. 1969).

Sebum production varies as a function of age and sex. It is low in children; in adults, it is higher in men than in women, and in men it falls only slightly with age, while in

Table 2.1. Various fatty acids

No. of carbon atoms	Fatty acid	Molecular formula	Melting point	Localization
2	Acetic acid	$C_2H_4O_2$	16.7°C	Vinegar
4	Butyric acid	$C_4H_8O_2$	−8°C	Butter
5	Isovaleric acid	$C_5H_{10}O_2$	−34.5°C	Dolphin oil
6	Caproic acid	$C_6H_{12}O_2$	3°C	Butter
				Coconut oil
8	Caprylic acid	$C_8H_{16}O_2$	−16°C	Butter
				Coconut oil
9	Pelargonic acid	$C_9H_{18}O_2$	15°C	Does not exist in natural state
10	Capric acid	$C_{10}H_{20}O_2$	31.5°C	Butter
				Coconut oil
16	Palmitic acid	$C_{16}H_{32}O_2$	63°–64°C	General animal oil
18	Stearic acid	$C_{18}H_{36}O_2$	71°C	Cow oil

Saturated fatty acids include acetic, butyric, caproic, caprylic, capric, lauric, formic, palmitic, and stearic acids, all of which contain carbon atoms. Saturated fatty acids have the general formula $C_nH_{2n}O_2$ and a specific individual odor

women it decreases significantly after they reach 50 years of age.

Sebum is enzymatically decomposed by normal cutaneous flora, *Corynebacterium acnes*, and *Staphylococcus albus*, a process that results in the rancid butter or cheesy odor associated with butyric and caproic short chain fatty acids, which are frequently detected when the follicle content is excreted (Hurley and Shelley 1960) (Table 2.1).

2.5   Hair

Human hair also provides sites that retain odorogenic sweat and make a substantial contribution to odor production since the hair accumulates axillary secretions, debris, keratin, and bacteria. Axillary odor is significantly reduced in shaved axillae, indicating that retained axillary secretion in combination with bacteria will increase the odor. If the axillae are regularly shaved, the odor is markedly diminished. Leyden et al. (1981) reported, however, that the presence of hair can contribute to odor intensity, but bacteria do not seem to be involved, acting only to the extent of enmeshing malodorous volatiles that have been generated on the skin surface as well as functioning as a disposal mechanism.

Axillary and pubic hair provide ornamentation as well as

erogenous stimulation. The continued presence of axillary hair after radical surgery for termination of bromidrosis is desired by male patients as a sign of masculinity.

2.5.1 Structure and Composition of the Hair Follicle

The structure and composition of the hair follicle are described by Baden (1987) as follows: a cylindrical depression formed by an invagination of the epidermis penetrates the corium into the connective tissue which contains the hair root. Attached to the follicles are sebaceous glands and the tiny muscles (erectores pili) which enable the hair to stand upright.

Hair is a complex tissue, with cells which are differentiated in various ways. The dermal papilla contains connective tissue and blood vessels, but the rest of the follicle is formed from modified epidermal cells, so that the hair shaft is produced from a multi-layered pilosebaceous unit.

At its lower end, the bulb is the thickest part of the follicle, and within it is a pool of undifferentiated cells (matrix) which proliferate and give rise to the various layers. Differentiation begins at the constriction site above the bulb. Protein synthesis continues up to the lower half of the follicle. At that point, the cells begin to expire and cornify in this keratogenous zone. Since the hair is cornified while still within the skin, it is completely hardened when it comes out of the follicle.

The outer root sheath, while contiguous with the superficial epidermis, has a different appearance at various levels in the canal. The next inward layer, the inner root sheath, is comprised of Henle's layer, then Huxley's layer, and the inner root sheath cuticle. Differentiation begins in Henle's layer and progresses inward. All of these three layers disintegrate within the follicle at the same approximate point as the entrance of the sebaceous gland. Since the inner root sheath is first to cornify, its main function may be to give shape to the hair (Fig. 2.10.a–c).

The next layer formed in the follicle is the cuticle. It is wrapped around the hair as it grows outward from the skin surface. The cuticle is composed of flattened, overlapping cuticle cells that show a laminar structure. It has two layers, the outer exocuticle and inner endocuticle. The cortex, which is the component of the hair fiber, is composed of elongated cells arranged parallel to the fiber axis. The medulla, which is the core portion of the hair, consists of

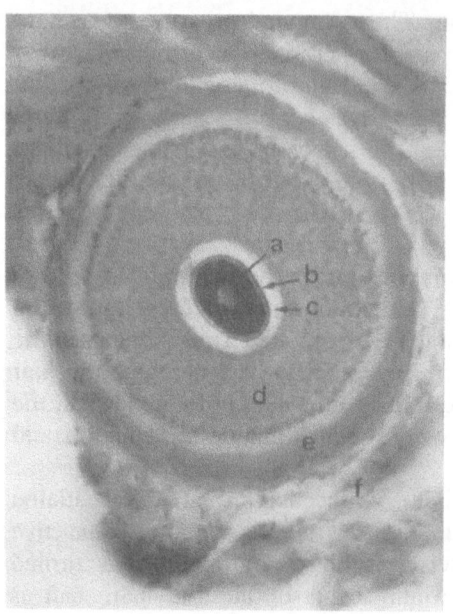

a

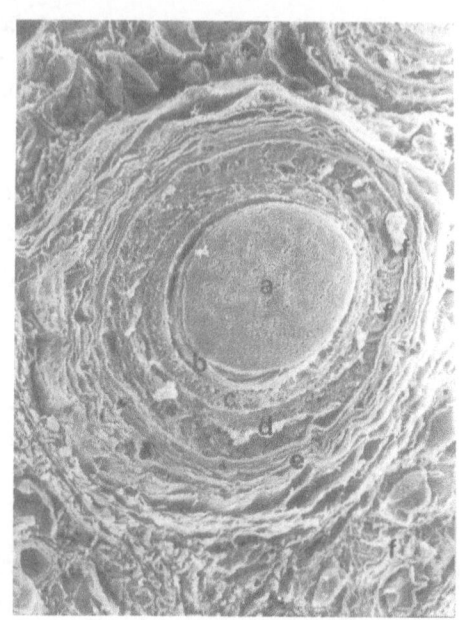

b

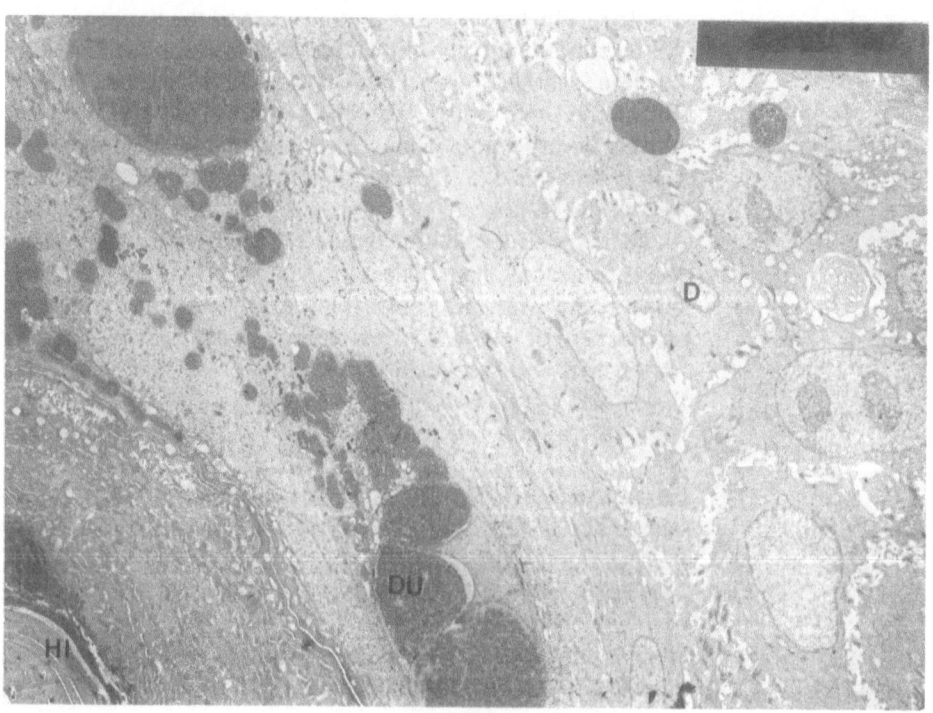

c

cells that have a unique type of differentiation. As these cells move upward from the papilla, spherical granules are formed, accumulated, and enlarged.

The dermal papilla functions as connective tissue in the hair bulb. Its size correlates with the size of the bulb, and it has a varying number of blood vessels, fibroblasts, ground substance, and fine collagen fibers. The hyaline membrane which surrounds the follicle is slender in its upper portion and thick in the lower third. It is enveloped by two collagen fiber layers in the follicle's lower tip: the outer one perpendicular to the long axis and the inner one running parallel. These connective tissue layers are continuous with the areola tissue which is wrapped around the sebaceous gland and the papillary layer of the dermis. This tissue is also connected to the dermal papilla through the stalk.

Blood vessels are believed to stem from the hypodermal arterial plexus and then go straight down to the dermal papilla's pore. Numerous shunts connect them around the follicle's lower third and form an abundant vascular plexus. No cross shunts can be seen in parallel vessels that lie above this zone to the sebaceous gland level where another plexus can be observed. Above this point, the parallel longitudinal vessels terminate in capillaries that envelop the follicle orifice.

2.5.2 Common Hair Cycle

The length of periodical fall-out and regrowth of hairs differs among various regions in the body. For scalp hair it is 5–7 years and for eyelashes and axillary hair much less (150 and 123 days, respectively).

The common hair cycle is divided into the three stages of anagen, catagen and telogen (Fig. 2.11). In the active or anagen stage, the dermal papilla, located at the base of the fully differentiated follicle, is almost wholly enclosed by matrix cells which exhibit rapid cell division. Daughter cells migrate upward to differentiate, forming the concentric layers of the hair (medulla, cortex and cuticle) and of the inner root sheath (cuticle, Huxley's layer, and Henle's layer) (Fig. 2.12.a). Mitosis in the hair matrix ceases during

←

Fig. 2.10.a–c. Cross section of the hair. **a** Histologic and **b** electrical scanning microscopic findings. *a*, hair cortex; *b*, cuticula; *c*, inner root sheath; *d*, outer root sheath; *e*, connective sheath; *f*, dermis. The sebaceous duct between the hair follicle and dermis. *HF*, hair follicle; *D*, dermis; *DU*, duct, ×7000

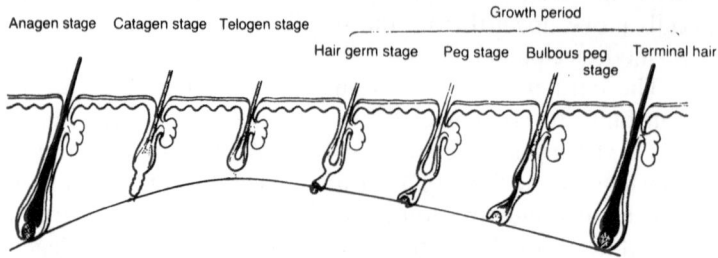

Fig. 2.11. The common hair cycle. (Reproduced with permission from Inaba 1985)

the catagen stage, and the base of the hair is then keratinized to form a brush or club hair. The hair follicle's lower
epidermal portion retreats from the dermal papilla to
shorten, and the hair bulb, contracting upward, forms a
club hair (Fig. 2.12.b,c). In the ensuing telogen stage, the
lower portion of the hair follicle is upwardly involuted, with
its volume reduced to about a half to a third the length of
an anagen follicle (Fig. 2.12.d).

The lower dermal papilla atrophies to a point located
close to the telogen follicle. Hair regeneration is observed
to start from the remnant cells of the dermal papilla at the
lower end of the telogen hair follicle (secondary hair germ)
which subtend the club end and the resting hair (Fig. 2.13.a).
The hair follicle pegs downward (hair peg) (Fig. 2.13.b),
after which the hair bulb, inner root sheaths and new hair
are formed from the hair matrix cell. As the newly-formed
hair moves upward, the club hair is pushed out and the new
anagen hair follicle is formed (Fig. 2.13.c–e). While the
hair cycles of human scalp hair and Japanese monkeys indicate the above findings, Moretti (1965) and Sato (1976)
state that typical telogen hair follicles are almost never seen
in scalp hair. Since the early anagen stage can be observed
in the lower portion of telogen hair follicles, the center of
hair regeneration was considered to be located at the lower
ends of telogen hair follicles. But if so, where does regeneration start after a telogen hair containing this supposed
hair center is pulled out?

Fig. 2.12.a–d. Hair follicle as observed in each stage of the hair
cycle. **a**, Anagen hair follicle, **b**, Catagen hair follicle, **c**, Enlargement of the lower portion of the catagen hair follicle, **d**, Telogen
hair follicle

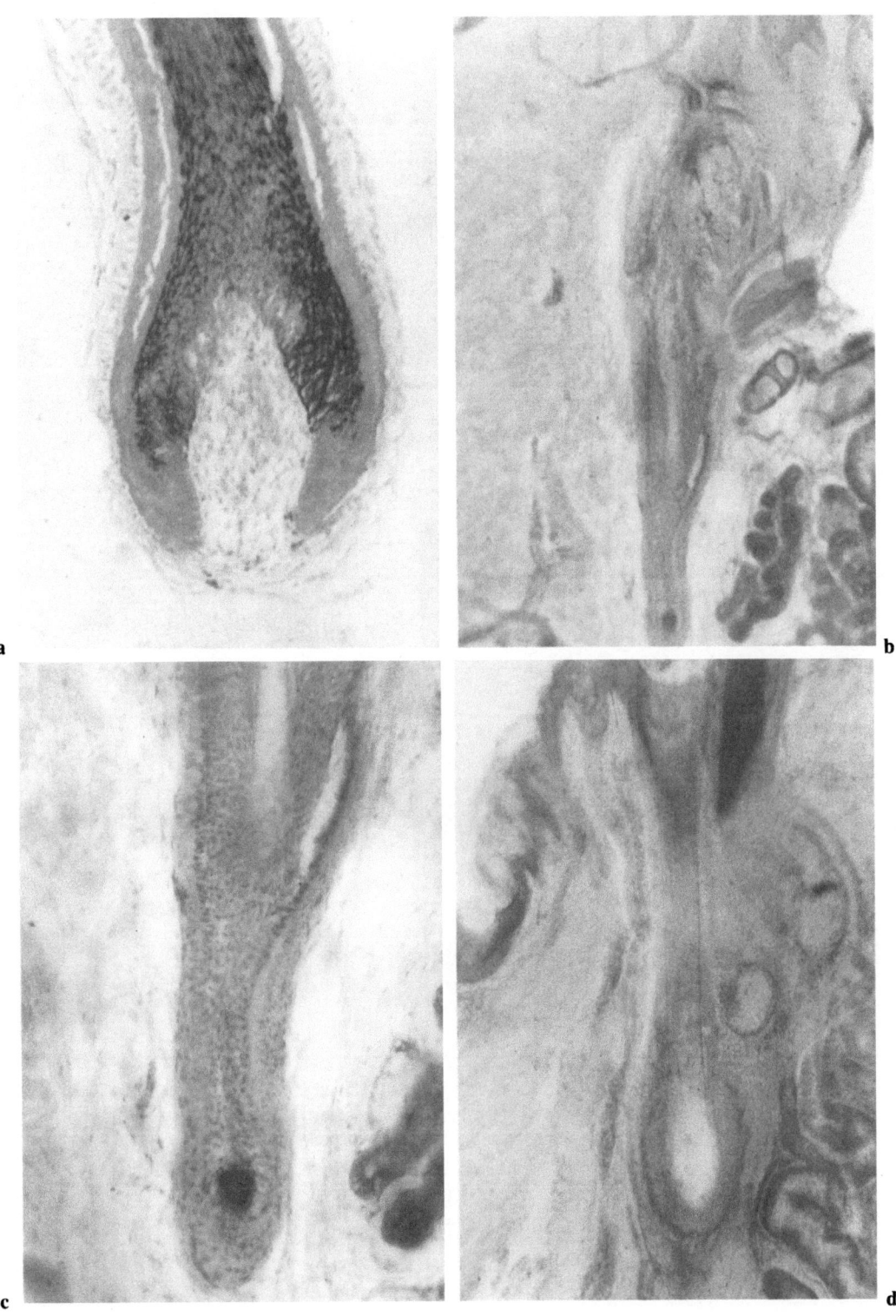

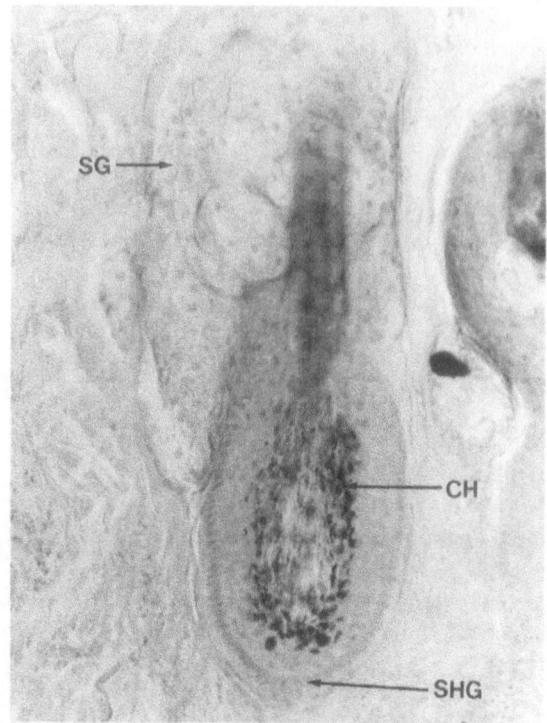

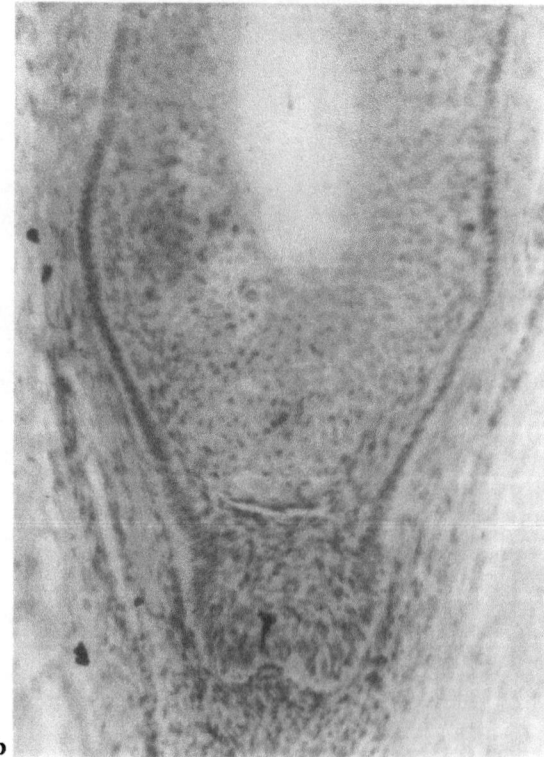

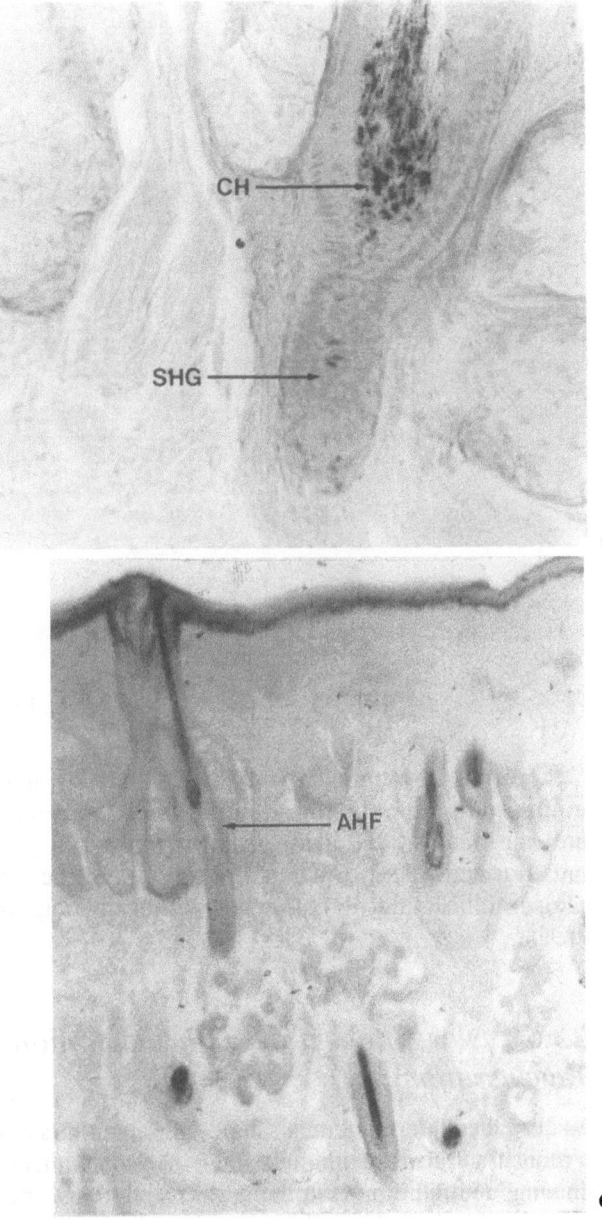

c

d

Fig. 2.13.a–e. Hair regeneration according to the common hair cycle. **a,** The secondary hair germ (*SHG*) starts to regenerate from the lower tip of the telogen hair follicle. *CH*, club hair; *SG*, sebaceous gland, **b,** Enlarged view of the more advanced stage of the regeneration of the secondary hair germ, **c,** The secondary hair germ (*SHG*) descends further downward. *CH*, club hair, **d,** The hair germ descends further to form the bulbous peg stage. *AHF*, anagen hair follicle, **e,** Enlarged view of the hair follicle in the bulbous peg stage. (*For Fig. 2.13e see following page*)

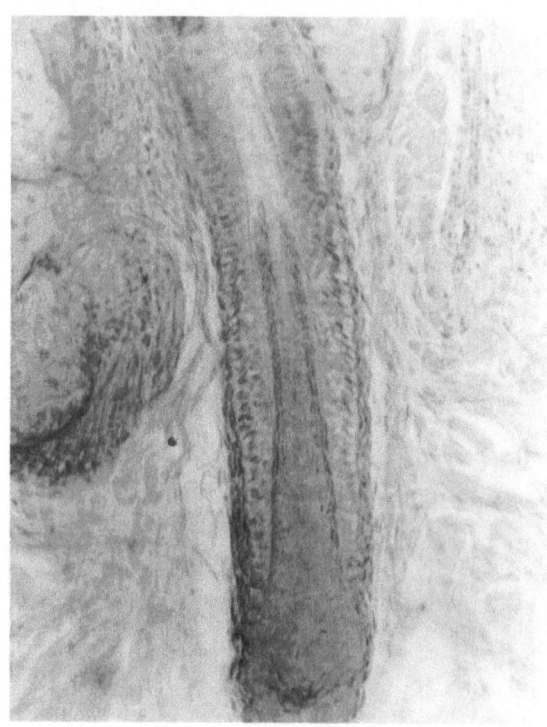

Fig. 2.13.e

According to Montagna and Parakkal (1974), hair regeneration starts from the lower end of the telogen hair follicle and can therefore be divided into two portions, permanent and transient. The former is located above the tip of the telogen follicle and the latter below that dividing line (Fig. 16.11).

2.5.3   New Concept of Hair Generation and Regeneration

As described later in detail (Chap. 15) the authors developed a radical surgical technique called the subcutaneous tissue shaving method for treatment of bromidrosis and hyperhidrosis. After applying this procedure, we raised a number of questions about the common concept of the generation and regeneration of the hair follicle (Inaba M and Inaba Y 1990) (Figs. 2.14.a,b; Chap. 16).

These findings indicate that the conventional hair cycle theory is incomplete, and that there is validity in our sebaceous gland hypothesis (Fig. 16.7.a–c) in which we propose that the true hair center is sited in the sebaceous gland and the follicle's upper isthmal portion. Taking this into ac-

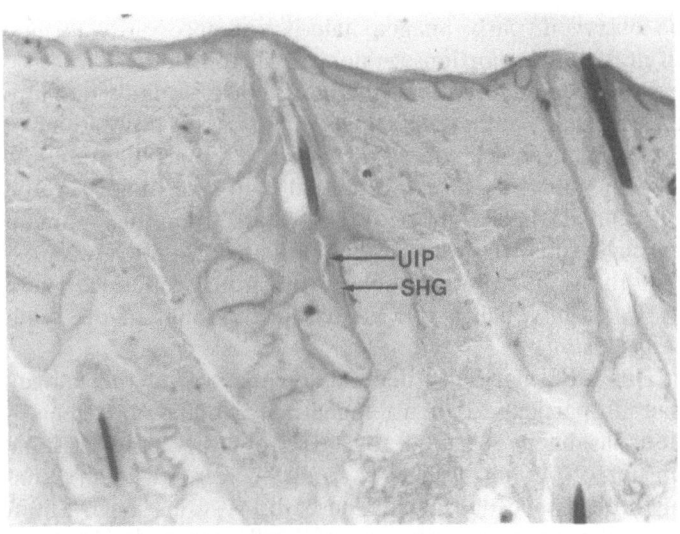

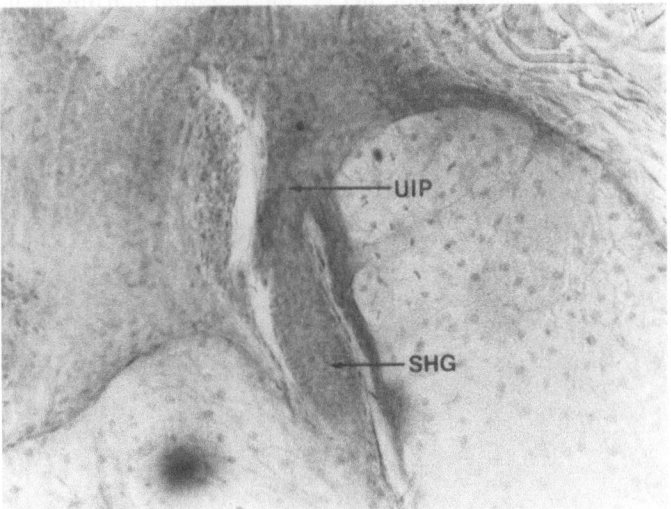

Fig. 2.14.a–b. Hair regeneration according to the essential hair cycle. **a** Secondary hair germ starts to regenerate from the upper isthmal portion after the telogen hair has been plucked. *UIP*, upper isthmal portion, *SHG*, secondary hair germ, **b** Enlarged view of the secondary hair germ follicle

count, we suggest that the hair cycle should be divided into four stages: anagen, catagen, telogen, and isthmal.

Formation of early anagen hair from the lower telogen hair follicles, which is consistent with the conventional hair cycle theory, has been observed in Japanese monkey and human scalp hair. However, since typical telogen hairs can be observed, early anagen hair is not necessarily formed from the lower portion of telogen hair follicles.

It has been thought that the hair follicle is subdivided into the transient and permanent portions at the lower end of the telogen follicle or at the point of attachment of the arrector muscle. The conventional hair cycle theory would be valid if regeneration were to start from this remnant dermal papilla. However, telogen hair does not always extend to the site of attachment of the arrector muscle. There are cases in which it retracts within the lobes of the sebaceous gland.

Thus, the hair germ depends on the size of the telogen hair. In particular, when telogen hair is epilated, regeneration has been observed to start from the upper isthmal portion of the hair follicle or from the secretory duct opening of the sebaceous gland. These observations suggest that, while the common hair cycle generally proceeds along the lines of the conventional theory, the essential starting point of regeneration is located in the upper isthmal portion (essential hair cycle) (Fig. 16.10).

Chapter 3. Mechanisms of Odor Development in Bromidrosis

It is not yet known whether the causative mechanism of bromidrosis is based on the density of apocrine and sebaceous glands or in the quality of gland secretory function. Much research has been conducted on the mechanism of odor development in bromidrosis.

3.1 Apocrine Glands

Results of research indicate that the odor depends on the quantity of sweat glands rather than their quality. In 1936, Ito reported no difference in apocrine gland quantities between normal individuals and patients with bromidrosis. Normal individuals also secrete small amounts of odoriferous material. In 1937, Adachi, however, stated after histological examinations that individuals with marked odor generally have far greater numbers of axillary apocrine sweat glands. He concluded that the degrees of individual odor depend largely on apocrine sweat gland development.

Even earlier, Honma (1925) reported on an iron-reactive substance detected in apocrine cell secretion. Moriyama (1927) stated that the difference in this iron-reactive substance between normal subjects and patients with bromidrosis was a very important factor.

A histological difference in the ceruminous glands of those who have wet, as opposed to dry, cerumen has been reported (Nagashima 1934; Hirayama 1942; Yoshihiro 1942). Abundant lipid droplets and pigment granules have been observed in the secretory cell cytoplasm of individuals with wet cerumen, but almost none in those with dry cerumen (Nagashima 1934). Secretory cells with striated cuticular borders were observed in almost all cells in cases of wet cerumen, but only in a few secretory cells if the cerumen was dry.

Apocrine sweat contains low unsaturated fatty acids. Its odor (mostly that of nonylic acid) was the same as that of bromidrosis cases (Nagamitsu 1941). A difference was detected in the amount of carbon (C) present in the fatty acids of the apocrine sweat gland in normal individuals (Nitta and Ikai 1954). The sweat has less than C4 lower fatty acids (acetic acid, propionic acid, formic acid) (Table 2.1), but in cases of osmidrosis the sweat has more than C4 lower fatty acids, for example, caproic acid. Differences in carbon number are responsible for the difference in odor between the osmidrotic and nonosmidrotic cases. In addition, ammonia constituents and volatile salts (trimethylamine, methylamine) may contribute to axillary odor. (The musky odor of sweat is attributed to saturated ketone and indole, and the odor of feces to fatty acids, indole, skatole, methyl mercaptan, hydrogen sulfide, etc.) Cases of bromidrosis reveal more apocrine gland fat, and the fat reacts with iron (Yoshihiro 1942).

Among other factors, bromidrosis patients show β-gluconidase activity in the apocrine glands and skin 10 times higher than normal.

3.2　Sebaceous Glands

Sebaceous gland secretion is known to be more abundant than apocrine gland secretion. According to Tamura (1915), the amount of axillary lipid in bromidrosis patients differed from the amount found in normal subjects. Different amounts of lipid and cholesterin ester and increased secretion of axillary fatty acid in bromidrosis patients have likewise been reported (Koyama 1925; Moriyama 1927).

Other reports (Thurnon and Ottenstein 1952) relate bromidrosis odor to ammonia quantity and amino acid, including sulfuric amino acid. Rothman (1954) reported that the odor originates not only in apocrine sweat but is also affected by sebum secretion from the sebaceous gland. Murata (1960) found that sebum influence makes the pH alkaline in cases of axillary bromidrosis. Subsequent bacterial contamination then gives rise to the acrid odor. Complex lipids secreted by sebaceous glands and epidermal lipids present on the skin surface have been found to reduce microbial populations (Ramasastry et al. 1970).

Skin surface lipids, degraded in part by gram-positive bacteria, form unsaturated fatty acids which are very active against gram-negative bacteria and certain fungi. On the other hand, some of those acids which are volatile might contribute to a strong odor.

Lipolytic activity in vitro has been observed in certain bacteria such as *Staphylococcus aureus* and *S. epidermidis*, in *Corynebacterium acnes*, and in other diphtheroids (Tagg et al. 1976).

Lipidic analysis has enabled studies of dermal surface lipid formation and composition. A correlation between lipid formation in sebaceous glands and on the skin surface has been frequently reported in cases of various dermatoses. Few studies have been done, however, on the dermal surface lipid of patients with bromidrosis.

It is clear that bromidrosis is caused chiefly by apocrine gland secretion. Eccrine sweat, sebaceous gland secretion, and bacterial activity are also implicated. Lipid formation in the axillary region has been clarified to some extent, but not fully.

We supplemented a previous comparative analysis of sebaceous gland lipid in patients with bromidrosis and normal controls with another study that investigated the dermal surface lipid of Japanese and Korean patients with bromidrosis for comparison with results obtained from normal controls (Inaba et al. 1987a).

Since it is difficult to obtain pure glandular sebum, most analyses rely on skin surface lipid, which is a mixture of sebaceous and epidermal lipids. Kellum (1966) reported that isolated sebaceous glands contain triglyceridous wax esters and squalene. Choi et al. (1984b) attempted to analyze triglycerides as the main component of sebaceous lipid and obtained the pure lipid. No difference was found between cases of bromidrosis with odor and those without odor (Chap. 6.1).

3.3   Mechanisms of Axillary Odor

The mechanisms responsible for axillary odor production have been gradually clarified. The most noted study on human axillary apocrine secretion was conducted by Shelley and Horvath (1951). They cannulated individual hair infundibulum to forestall axillary eccrine secretion and bacterial contamination. In a paper which summarized their findings (Shelley et al. 1953), they reported that apocrine sweat is sterile and odorless when it first appears on the skin surface. Axillary microorganisms which act on apocrine sweat in vitro promote development of the typical acrid axillary body odor within a few hours. If these organisms are excluded from sterile apocrine sweat or if their growth is inhibited by adding hexachlorophene to the sweat, the odor will not develop. Apocrine sweat, they noted, is the only substrate required to produce the odor. They further reported that the odor can be prevented for more than 18 h in most people by intensive use of hexachlorophene-detergent preparations.

Axillary hair substantially increases the odor, since the hair accumulates axillary secretions, debris, keratin, and bacteria. However, shaving and scrubbing the axilla can abolish odor for more than 24 h.

Pure eccrine sweat, sterile or unsterile, has no odor and will not develop any. Only when contaminated with sebum keratin or debris does eccrine sweat develop an odor as a result of bacterial action. This odor is mild and quite distinct from the odor developed in pure apocrine sweat. In association with the production of axillary odor, eccrine sweat accentuates bacterial growth and participates in volatilizing odoriferous compounds derived from apocrine sweat.

Shelley found no significant reduction in eccrine or apocrine sweating after the application of aluminum salt preparations. These preparations, however, had strong deodorant activity. In daily application, they can change or eliminate typical axillary odor, in most cases, for 12–18 h. With use of aluminum chloride, this deodorant activity is based on antibacterial action and has a chemical effect on odoriferous products. Subsequent studies (Takami 1960; Kobori and Narumi 1958) provided further evidence, more or less, that Shelley's conclusions were correct.

3.4   Bacterial Activity and Skin Odor

Kobori agreed with Shelley up to a point, but then insisted that patients with bromidrosis have specific apocrine sweat materials that produce odor due to bacterial degradation. Normal subjects do not exhibit that much apocrine secretion. He reported a gram-positive *Staphylococcus albus* coryne form in almost all specimens examined. Slight populations of *Staphylococcus citreus* and *streptococcus* were also observed. Shehadeh and Kligman (1963) reported only gram-positive species, most notably coagulase-negative *staphylococci*. Takami (1960) found no difference in bacterial populations among patients with bromidrosis and those without. The bacteria present were mostly coagulase negative *Staphylococcus albus*, although, in a small number of specimens, he also found gram-negative *Staphylococcous citreus*.

As a vital body organ, the skin has numerous functions. One of them is the control of an overgrowth of normal

Table 3.1. Bacteria in the axillary region

No. of patients	Bacillus	Staphylococcus Coagulase-negative	Coagulase-positive	Proteus	Candida	Yeast fungus	Hay bacillus	Total no. of germ carriers	Total no. of germ-free patients
19		4	1	3	1	1	1	11	8

(Reproduced with permission from Inaba 1986)

microbial flora. This is very significant because bacterial metabolism is implicated in odor production. Apart from functioning as a mechanical barrier, the skin has a low pH (5.0–6.5) and continually sheds epidermal scales. Both of those functions contribute to control of bacterial populations (Youmans et al. 1975).

We also examined the presence of bacteria in the axillary region (Table 3.1) (Inaba 1986). Among 19 patients, 11 had identifiable bacteria; the others had no significant bacteria numbers. *Staphylococcus* was detected in five cases (45%). Almost all of the bacteria detected in four cases were coagulase-negative *staphylococci* and in one case, coagulase-positive *staphylococci*. The rest revealed various bacteria present in the axillary hair complement (*Proteus, Candida,* yeast fungus, and *Bacillus subtilis*). Bacteria attached to the hair shaft consisted only of coagulase-negative *staphylococci* (white *Staphylococcus [albus]*).

Labows et al. (1982) reported a study of axillary flora in both male and female subjects in which the microflora was quantitively stable over the period of observation. Males and females had approximately the same number of organisms per square centimeter. Research into bacteria from the axillary hair shaft in five patients revealed that they all had *Staphylococcus* coagulase-negative and none had *Staphylococcus* coagulase-positive. Microflora composition is shown in Table 3.2.

Comparatively dense populations of aerobic cocci (mainly *Staphylococcus epidermidis*) were found in all subjects and did not have a decisive role in the quality of axillary odor. A significant difference in populations of lipophilic diphtheroids was observed between males (85%) and females

Table 3.2. Prevalence and density of axillary-resident bacterial flora in males and females

	Males (n = 128)		Females (n = 77)	
	%	Density	%	Density
Total Aerobes	100	5.84	100	5.95
Micrococcaceae	100	5.51	100	5.56
Lipophilic diphtheroids	85	5.40	66	5.36
Large colony diphtheroids	26	4.43	25	4.57
Gram-negative rods	20	3.36	19	3.32
Total Propionibacteria	70	3.71	47	4.23
P. acnes	47	3.86	30	4.26
P. avidum	34	3.62	21	4.18
P. granulosum	8	3.61	5	3.65

(Reproduced with permission from Labows et al. 1982)

(66%). The other category of diphtheroids was less abundant, approximately 25% in both males and females. About 20% of the subjects had various gram-negative (*Escherichia*, *Klebsiella*, *Proteus*, *Enterobacter*) populations, but in low quantities averaging around $1000/cm^2$ and not contributing significantly to apocrine odor production. Another sex difference was found in the presence of anaerobes (*Propionibacterium* species) in males (70%) and females (47%). However, anaerobes do not participate in odor production since they reside in follicular pore depths in which oxygen tension is low.

3.5   Recent Research

The opening of the human apocrine sweat gland on the skin surface is shared with the hair follicle, which makes it difficult to ascertain if the secretion reported by Shelley (1951) was in fact a mixture of sweat gland and sebaceous gland secretions.

Labows (1979) describes human axillary odor as the product of microbes acting on an essentially odorless secretion. These microbes are mainly lipophilic diphtheroids and micrococci bacteria. The latter produce an odor characteristic of isovaleric acid. This same acid is also produced by the diphtheroids, but the odor is overwhelmed by other odor components, which give rise to a stronger apocrine odor. Proliferation and odor of lipophilic diphtheroids are increased by sebum.

Androstene, androstenediol, and other steroids have also been found in trace amounts in axillary sweat. It is evident that heated apocrine secretion releases an apocrine-like odor attributed mostly to isomeric androstadien-17-one and androst-2-en-17-one produced by dehydroepiandrosterone and androsterone sulfates, respectively, in thermal breakdown. These findings make it clear that apocrine secretion includes specific steroid substances as well as cholesterol which diphtheroid bacteria may metabolize to odorous 16-androgens.

According to Leyden et al. (1981) common axillary odor is generated by aerobic diphtheroids, and incubating micrococci with apocrine sweat produce an odor akin to that of isovaleric acid. Short-chain fatty acids and ammonia are the major odor components produced by apocrine sweat that has been degraded by bacteria, but other unidentified substances may be present as well. That could explain the various odors, ranging from sweaty, rancid, and fecal to sour and sweet, that are sometimes detected, indicating individual variations in the chemical composition of axillary perspiration.

According to Kanda et al. (1990) short-chain fatty acids from shirts in the axillary region were extracted with ethyl ether, and then analysed by gas chromatography/mass spectrometry (GC/MS). As a result, low-grade fatty acids of C6, C8, C9, and C10 were detected from the shirts in the axillary region although the analysis data varied slightly according to each individual. C9 (pelargonic acid) and C10 (capric acid) were detected from all subjects; however, no low-grade fatty acids were detected from the regions other than the axillae.

Certain androgenic steroids, specifically dehydroepiandrosterone sulfate and androsterone sulfate, have odors similar to natural axillary odor, as do androst-16-en-e-one and androst-16-en-3-ol. One or more of them may be the apocrine sweat odorogens. However, none of them have been identified as present in axillary secretions. The distinctive apocrine odor comes only from the axilla. Other apocrine sites, for example, the pubic area, contain the bacterial flora as well, but the amount of apocrine sweat is insignificant. The apocrine glands at those sites may even be functionless (Kligman and Shehadeh 1964).

The common apocrine odor is produced only by bacterial degradation of apocrine sweat. It cannot be ascribed to sebum, hair, eccrine sweat, or keratin scales, singly or in any combination. The odor is so intense that even though it is localized only in the axillary fossae, it is called "body odor."

3.6   Methods of Odor Detection

The proper study of odor sources in human skin requires the examination of uncontaminated samples of secretion or skin component. Moreover, a distinction must be made between odors intrinsic to the specimen and those which develop secondary to degradation or chemical alteration, usually by bacteria. Except for careful studies of apocrine odor, investigation of other cutaneous odors has not been properly critical. In addition to apocrine sweat, possible sources of odor from human skin include eccrine sweat, hair, nails, and keratin scales.

Present analytical methods such as headspace concentration, gas chromatography, and the combination of gas chromatography/mass spectrometry (GC/MS), have made it possible to separate and identify submicrogram quantities of organic compounds (Dravnieks 1975; Labows et al. 1979b).

Total body volatiles have been sampled by placing individuals in glass tubes and sweeping with air to concentrate the odors (Dravnieks 1975). A telephone booth-like chamber was also used to sample human volatiles, and over 100 chemicals were identified (Ellis et al. 1974). These experiments examined the possibility that body odors might be unique enough to serve as personal signatures. Techniques for the concentration of vaginal, oral, skin, and axillary odors have also been developed (Dravnieks 1975; Labows et al., 1979b). In addition, devices have been made for monitoring skin volatiles for mosquito attractants (Price et al. 1979). The sampling and identification of volatiles from different body sites and their application for the diagnosis of disease has recently been reviewed (Sastry et al. 1980).

The approach to the study of skin odors has been to duplicate the natural odors in vitro by incubating the resident bacteria with the appropriate skin secretion. The contribution of the scalp yeast, *Pityrosporum ovale*, to scalp odor has been determined by this approach (Labows 1979).

GC/MS profiling of the small organic compounds present in body secretions, such as blood serum, cerebrospinal fluid, and urine of diseased and healthy individuals, provides useful diagnostic information. In some cases, complete metabolic profiles are obtained and qualitative or quantitative changes in individual components are noted with the onset of disease processes or throughout the female reproductive cycle.

In the identification of the chemicals responsible for skin odors, or for the odors produced in cultures by microorganisms, GC/olfactory analysis is used to determine which components of these complex mixtures contribute to the observed odor. However, identification of the chemical nature of odorants is now almost always feasible, using gas chromatographic and spectrometric methods. Odor detection, at least in medical diagnosis, is dependent upon the unaided nose of the examiner, although specialized and cumbersome olfactometric methods are available. The olfactory limitations of an occasional individual—whether they be transient or permanent, of broad range or limited to the perception of a single primary odor—should be recognized.

Chapter 4. Relationship Between Hyperhidrosis and Bromidrosis

Emotional hyperhidrosis (EH) is ordinarily localized in the palms, soles, and axillae. In some cases it may arise from a systemic disease, but it can also occur as a primary idiopathic process. It is one of the localized forms of idiopathic hyperhidrosis. Those individuals who are prone to EH may be embarrassed and socially disadvantaged. Emotional hyperhidrosis, especially in the axillary area, involves only the eccrine glands, and the sweat does not stain clothing in spite of soaking it. On the other hand, axillary bromidrosis is almost always accompanied by a discolorative apocrine hyperhidrosis.

As stated previously, there is a close relationship between bromidrosis and the amount of moisture in cerumen. In treating Japanese patients, the authors divide cases of hyperhidrosis into two types:

1. Essential hyperhidrosis—hyperhidrosis characteristic of individuals who have dry cerumen (palms, soles, and axillae)
2. Symptomatic hyperhidrosis—hyperhidrosis characteristic of those who have wet cerumen (axillary region)

Since bromidrosis is almost always accompanied by hyperhidrosis, those cases characterized by wet cerumen are classified as having symptomatic hyperhidrosis. About 70% of patients with wet cerumen have hyperhidrosis combined with odor, which is characteristic of bromidrosis. About 30% of the remaining patients with wet cerumen, however, have excessive numbers of eccrine glands but only small numbers of apocrine glands. In those cases, there is no odor, only hyperhidrosis such as that characteristic of symptomatic hyperhidrosis.

4.1 Mechanisms of Sweating

Emotional hyperhidrosis (cortical sweating) differs from the
thermoregulatory type. Emotion-generated eccrine sweating
may appear in response to intense emotional stimulation.
This type of hyperhidrosis is common in all races and in
both sexes. Thermoregulatory (hypothalamic) sweating is of
a different order and is produced in one of two ways: by the
increased temperature that directly affects the center heat in
the anterior hypothalamus, or by a reflex from the stimulated nerve endings of the skin, not necessarily those which
transmit the sensation of heat. The hypothalamus is a major
autonomic center within the central nervous system. As
such, it contains cells that trigger impulses which regulate
thermoregulatory sweating as an essential physiological
eccrine response.

Rapid evaporation and cooling are characteristics of heat-
responsive sweating. A rise in blood temperature will stimulate the hypothalamic center. A sympathetic discharge
derived from vasomotor disorders may set off essential
hyperhidrosis due to the synchronous discharge of vasoconstrictor and sudomotor impulses that result in a "cold
sweat." Adrenergic sweating is produced by epinephrine,
the excessive perspiration of patients during paroxysmal
attacks of pheochromocytoma, and possibly the sweating of
hyperthyroid patients, although this may be partly due to
a thermoregulatory mechanism. It can be suppressed by
adrenolytic agents such as Dibenamine (N,N-dibenzyl-β-
chloroethylamine) (Takatsu 1957).

Other cases of perspiration related to essential hyperhidrosis or bromidrosis include gustatory sweating in which
salivation is produced by drinking, eating, or even chewing,
as a characteristic of the auriculotemporal syndrome (Frey
1923). In cases of compensatory sweating, the sweat glands
become hyperactive in a specific skin site as a result of
hyperactivity in another site. Kuno (1956) has discussed
important differences observed between cortical and hypothalamic sweating. In a cold or temperate climate, for instance, the thermoregulatory center (hypothalamic
center) is ordinarily dormant due to a low rate of excita-
tion, but will react increasingly to a high rise in local
temperature.

By contrast, the cortical sweat mechanism, which is re-
active even at normal room temperature, will react immediately to an appropriate impulse. Emotional stimuli can
readily set off emotional sweating which has no latent phase
such as that observed in hypothalamic sweating.

4.2   Essential Hyperhidrosis

This is a severe form of hyperhidrosis which may affect all aspects of the patient's life and influence social behavior and professional development, although it is not a threat to life or organ integrity. Diagnosis of essential hyperhidrosis (EH) is determined by clinical observation of sweating palms or large, underarm wet spots on clothing with no discoloration. The patients may have experienced excessive sweating since birth, or at least since early childhood, and there is often a familial history. It seldom occurs for the first time after 20 years of age. Many patients notice the condition beginning with the onset of puberty.

These patients, frequently female, are otherwise healthy adolescents with no overt physical disorders. Their history is one of excessive axillary and/or volar sweating which can be provoked by psychological stress in meeting strangers, taking school examinations, or during the performance of delicate physical movements involved in writing or sketching. If sweating is repeatedly excessive, the hands may turn beefy red, or the feet and axillae become macerated and secondarily infected. Sweating is minimal during sleep or the induction of anesthesia. It is thus likely that the cause of EH is **an excessive** outflow of centrally originated sudomotor **stimuli.**

Examination of sympathetic nervous tissue has, with only one exception, revealed no abnormality in the histological appearance of the sympathetic ganglia. This condition may be based on abnormally high numbers of sweat glands and their degree of reactivity. Human skin is commonly moist to some extent due to eccrine sweat and normal epidermal evaporation. One study of young people in Israel yielded a 0.6%–1% incidence of hyperhidrosis of all areas and severity. One-fourth of these cases indicate severe palmar hyperhidrosis (Adar et al. 1977). Race may be a factor in hyperhidrosis incidence, according to Cloward (1969), who found a high frequency in patients of Japanese ancestry.

Essential hyperhidrosis seems to be hereditary. Cloward found that 19 (23%) of 82 patients with EH had one or more EH relatives. Adar et al. (1977) reported that 53 out of 100 patients had EH in their family history, and 21 of them were found to have family histories that revealed frequent cases of severe palmar hyperhidrosis among first-degree relatives. Shih and Wang (1978) reported 63 out of 264 patients as having a similar family history with pronounced hyperhidrosis in first-degree relatives. Solomons (1962), in a study of 12 families with hyperhidrosis, concluded that the condition was transmitted by an autosomal-dominant gene. Other studies, however, do not reveal a specific predisposition. In cases of true bromidrosis, the incidence of this specific hyperhidrosis condition is evidently less than 0.5%.

The symptom of palmar hyperhidrosis is psychological sweating, and patients are tormented by excessive sweating resulting from a psychological obsession. Hand shaking is not a custom in Japan, which is a relief for those afflicted with palmar hyperhidrosis, but they are nevertheless continuously obsessed with the idea that they should not hold out their sweaty hands for a handshake, which may hinder a smooth personal relationship. Apart from hand shaking, writing is another big problem for the patients when papers get wet by moisture secreted from their palms. To lessen their anxiety, the practice of using a brush may be one of the solutions, since hands do not touch the paper in this form of writing. Such solutions may surely lead to a psychological relief and help build up confidence for better personal relations.

Chapter 5.   Relationship Between Cerumen and Bromidrosis

There are two distinct types of normal human cerumen. The usual type of cerumen found among Japanese, Chinese, Mongolian, and American Indian populations is brownish-gray, brittle, and dry in nature. It is commonly called "rice-bran cerumen" (*nuka-mimi*) in Japan because it resembles a particle of rice bran. When accumulated in the external auditory meatus for a long period of time, a thin slice of this cerumen of a fairly large size will often form. The other type of cerumen is brown, sticky and wet. This type is sometimes called "honey cerumen" (*ame-mimi*) or "cat cerumen" (*neko-mimi*). The wet type is very prevalent among European and African populations. An intermediate type which can be classified as neither wet nor dry occurs at a low rate of about 0.5% among healthy Japanese (Matsunaga 1962).

Normal cerumen is a mixture of the secretory products of ceruminous glands and sebaceous glands distributed in the external auditory meatus in addition to exfoliated epithelial cells. The difference between the wet and dry cerumen types is attributed to the difference in the secretory products of the ceruminous glands (the so-called ceruminous apocrine glands). On the other hand, axillary bromidrosis is due to the condition of hormonally influenced apocrine sweat glands. Although full development of these apocrine glands in the axilla does not begin until puberty (under hormonal influence), their development in the external auditory meatus begins immediately after birth.

5.1 Frequency of Wet Cerumen and Bromidrosis

Since there is a close correlation between the condition of cerumen and bromidrosis, a number of papers have reported the incidence of wet cerumen in the Japanese population. The results obtained by previous workers are summarized in Table 5.1 according to Matsunaga (1962), who reported that frequencies of wet cerumen in different parts of Japan vary within the range of 12.6%–22.4% and, to some extent, are probably due to local differences (Matsunaga et al. 1954). No difference in the frequencies of wet and dry types was detected between the sexes (Matsunaga 1962). In a more recent study, Petrakis et al. (1986) reported on the frequency of dry cerumen in European populations, and these results are shown in Table 5.2.

5.2 Inheritance of Cerumen Types

A number of family studies have been published. Miyake (1932) reported that wet cerumen, if present in one Japanese parent, is inherited in a proportion of 52.23%. If both parents have dry cerumen, the children do not have wet

Table 5.1. Frequency of wet cerumen in Japanese populations

Place and population	No. examined	Frequency of wet cerumen		Authors
		No.	%	
Tokyo, college students and nurses	1,893	273	14.42	Nagashima (1934)
Kyoto, healthy	432	81	18.75 ⎫	Nagai (1935)
Kyoto, insane	492	88	17.88 ⎭	
Age: 15–59	7,816	1,255	16.06	Adachi (1937)
Amakusa island	419	85	20.29 ⎫	
Kumamoto	611	95	15.55 ⎬	Kutsuna (1939)
Fukuoka	405	69	17.04 ⎪	
Kagoshima	208	19	9.13 ⎭	
Oita	416	73	17.55	Amako (1939)
Seoul, Japanese school children	2,813	355	12.62	Nozoe (1943)
Hokuriku district	831	185	22.26	Morita (1947)
Kyoto	1,018	179	17.58	Ichida et al. (1949)
Yokosuka	380	70	18.42	Takamiya (1952)
Tokyo, age: above 20	642	144	22.43 ⎫	Matsunaga et al. (1954a)
Sapporo, pupils of a middle school	1,638	311	18.99 ⎭	
Tokyo	3,403	528	15.52	Hayashi (1958)
	23,417	3,810	16.27	Total

(Reproduced with permission from Matsunaga 1962)

cerumen. Other reports by Adachi (1937), Suzuki (1938), Nozoe (1943), Morita (1947), Ichida et al. (1949), and Takamiya (1952) show results in general agreement with the supposition that the wet cerumen type is controlled by a pair of completely dominant autosomal alleles. Matsunaga (1962) demonstrated that cerumen type is derived from simple Mendelian inheritance. The wet type is dominant homozygous (W × W) or heterozygous (W × w). The dry type is homozygous recessive (w × w). The heterozygous wet type (W × w) cannot be clinically differentiated from the homozygous dominant type (Table 5.3).

Several studies have reported no change due to aging in proportions of wet and dry cerumen types. (e.g., Nakajima and Hirano 1968, Hayashi 1958). This typing has also been used as an index to establish parental relation in forensic medicine. On the other hand, Petrakis et al. (1986) reported a study of white females that showed a notable increase in the dry phenotype with advancing age.

Matsunaga (1962) reported Japanese family data which indicate single factor inheritance of cerumen types. A

Table 5.2. Frequency of dry cerumen in European populations

Population	Number tested	% Dry	Frequency of dry allele (q)	Reference
German	514	3.10	0.176	Matsunaga and Ebbing (1956)
English	125	2.25	0.150	Petrakis (unpublished)
Icelandic	322	1.20	0.111	Lehmann et al.
Finnish (NE)	323	2.50	0.157	Eriksson et al. (1980)
Spanish—Basque	491	5.09	0.225	Calderon (1977)
Castilian	308	2.27	0.151	Calderon (1977)
Andalusian	209	0.96	0.098	Calderon (1977)
Greek	81	6.17	0.248	Spanidu and Petrakis (unpublished)

(Reproduced with permission from Petrakis et al. 1986)

Table 5.3. Japanese family data indicating single factor inheritance of cerumen types

	Mating			Children					
	No.				Wet		Dry		
Type	Obs.	Exp.	χ^2	Total	Obs.	Exp.	Obs.	Exp.	χ^2
Wet × Wet	14	16.1	0.27	47	35	36.7	12	10.3	0.36
Wet × Dry	116	111.7	0.17	400	205	212.7	195	187.3	0.60
Dry × Dry	191	193.2	0.03	634	0	0	634	634.0	
Total	321	321.0	0.47	1081	240	249.4	841	831.6	0.96

d.f. = 2, $0.80 > P > 0.50$
(Reproduced with permission from Matsunaga 1962)

present-day example of family data, comprising 321 families with 1081 children collected in Tokyo, is presented in Table 5.3. If both parents have dry cerumen, wet cerumen was not detected in any of their 634 children. Since observation and expectation are in close agreement, the result endorses the above hypothesis.

Nakajima and Hirano (1968) reported similar findings (Table 5.4). Cerumen types were studied in a sample of 1071 individuals residing in one provincial town of Japan. Frequency of wet and dry types was 19.9% ± 1.2% and 80.1% ± 1.2%, respectively, for the whole sample, and 19% ± 2% and 81% ± 2%, respectively, for 370 unrelated parents. No significant differences were found in the frequency between the sexes and among age groups.

Differentiation of cerumen types among children and the proportion of families without a dry-type child show consistent agreement with the supposition of simple Mendelian inheritance; the dry type is an autosomal recessive character and the wet type a dominant one (Table 5.5).

We reported on the frequency of bromidrosis in Japanese families (Inaba et al. 1975). Patients were questioned about body odor in the family. Since the odor is perceived differently by each individual, the answers were no less correct than if given in response to direct inquiries about the condition of the cerumen.

Table 5.4. Frequency of cerumen type by sex

Cerumen	Male		Female		Both sexes	
type	No.	%	No.	%	No.	%
Wet	113	20.85	100	18.90	213	19.89
Dry	429	79.15	429	81.10	858	80.11
Total	542	100.00	529	100.00	1071	100.00

$\chi^2 = 0.119$, d.f. = 1, $0.9 < P < 0.95$
(Reproduced with permission from Nakajima and Hirano 1968)

Table 5.5. Test for the hypothesis of simple Mendelian inheritance

Mating		No. of families	Total no. of children	No. of wet-type children		No. of dry-type children		χ^2
Father	Mother			Observed	Expected	Observed	Expected	
Wet	Wet	8	30	27	23.257	3	6.742	2.680
Wet	Dry	25	117	64	61.562	53	55.437	0.204
Dry	Wet	32	129	73	67.824	56	61.176	0.833
Dry	Dry	120	345	0	0	345	345.000	0.000
Total		185	620	163	152.643	457	468.355	3.717

$\chi^2 = 3.717$, d.f. = 3, $0.25 < P < 0.3$
(Reproduced with permission from Nakajima and Hirano 1968)

As shown in Table 5.6, the inheritance was from the paternal line in 124 cases (31%) and from the maternal line in 108 cases (27%). Both parents were involved in 16 cases (4%), and siblings had the same odor in 41 cases (10%). A total of 289 patients (72%) had a clear hereditary factor of bromidrosis, but 111 (27%) may be sporadic cases in which neither parents nor siblings have axillary odor. Among them, 48 families had only one child (12%), and in 63 families (16%) patients were utterly obsessed with the notion that they are the only ones among their family members having a predisposition to bromidrosis. However, as will be explained later, it has been found that the determination of a predispositional tendency should depend on whether the patient has dry or wet cerumen. Yoshihiro in 1942 and Takami in 1960 reported similar results.

In conclusion, we assessed the sporadic cases in which the parent did not have any odor in spite of having wet cerumen (symptomatic hyperhidrosis), i.e., the odor was undetected due to regular use of a deodorant, mutation, and/or other factors. It was not possible to make a further determination of these 111 sporadic cases without further research among the total of 400 cases. An attempt was made to research the hereditary factor only in obvious cases, setting aside those sporadic cases in which only one sibling had the same condition, which constituted a minority (Table 5.6). In the final analysis, the frequency of only one parent having bromidrosis was 55%, while both parents having bromidrosis was 81% (Table 5.7).

In a more recent study (Inaba, unpublished work), we again examined this hereditary factor. In 124 families

Table 5.6. Frequency of bromidrosis in Japan

| | Yoshihiro 1942 | | Takami 1960 | | Inaba/Nishida report 1973 | | Average |
	No.	%	No.	%	No.	%	%
One parent Paternal line	55	30	13	46	124	31	36
One parent Maternal line	50	27	8	29	108	27	27
Both parents	11	6			16	4	5
Sibling + Parents −	26	14	6	21	41	10	15
Morbidity of family	142	78	27	96	289	72	82
Sporadic case	41	22	1	4	111	28	18
Total	183	100	28	100	400	100	

289 patients (72%) had a clear hereditary factor of bromidrosis, but 111 cases (28%) may be sporadic cases. (Reproduced with permission from Inaba 1986)

Table 5.7. Frequency of the hereditary factor

	No. of families	No. of children		Bromidrosis		Nonbromidrosis		χ^2	d.f.	P
				No.	%	No.	%			
Both parents have bromidrosis	12	43	No. of cases	35	81	8	19	0.616	1	0.50 > P > 0.30
			Expected no.	32.81		10.19				
One parent has bromidrosis	157	557	No. of cases	307	55	250	45	3.235	1	0.10 > P > 0.05
			Expected no.	285.78		271.22				

(Reproduced with permission from Inaba 1986)

(D × W) of 321 children, wet cerumen was confirmed in 175 children (54%) and dry cerumen in 146 (46%). In 7 families (W × W) with 19 children, 15 (79%) had wet cerumen, and 4 (21%) had dry cerumen.

5.3   Relationship Between Bromidrosis and Wet Cerumen

Inaba et al. (1975) reported that almost all patients with bromidrosis have wet cerumen. Nagai (1935) reported bromidrosis in 74.1% of patients with wet cerumen. Takami (1960) reported a ratio of 94% in males and 95.4% in females. In individuals with wet cerumen, the axillary glands are usually more developed than in those with dry cerumen. However, it is not likely that every individual with wet cerumen gives off a strong odor. According to results reported by Nagashima (1934), who examined 1893 persons of ages 16–30 years, 211 out of 273 cases of wet cerumen (i.e., about 77.3% of the wet type) were observed to have axillary odor. Hirota (1939) reports 91.4% and Yoshihiro (1942) 78.8%. The rate of frequency of this condition increases with age more in females than in males.

Adachi (1937) reported on an extensive study of racial differences in axillary odor in connection with cerumen types. The results showed that there is a marked racial difference in the relative frequencies of wet types among different ethnic groups: Northern Chinese, 4.2%; Koreans, 7.6%; Micronesians, 62.9%; Formosan aborigines, 71.4%; and Ainu aborigines, 86.7%.

Suzuki Y (1938) and Suzuki A (1960) reported an incidence of 12.8%–15.7% in the Japanese population. Nozoe (1943) reported 12.6% in Japanese school children residing in Seoul. Morita (1947) reported 22.3% in the Hokuriku district, Ichida et al. (1949) 17.6% in Kyoto, Takamiya (1952) 18.4% in Yokosuka, and Matsunaga et al. (1954) 15.5% in Tokyo. The mean frequency for the pooled data of about 23,000 individuals examined is 16.3% (Matsunaga 1962).

Few patients with dry cerumen are found to suffer from bromidrosis. Takami (1960) reports a low 5% in males and 2.7% in females. Yoshihiro (1942) reports 2.9%. We detected 17 cases of bromidrosis among 350 patients (5%) with dry cerumen, the commonest being those of pubertal age with bromidrosis occurring because of strong hormonal secretions. Cases were also divided into two types: (1) essential hyperhidrosis with dry cerumen, and (2) sym-

ptomatic hyperhidrosis with wet cerumen, or the constitution of bromidrosis in the broad sense.

Although development of the apocrine glands in the
axilla does not begin until puberty under hormonal influence, the development of these glands in the external
auditory meatus begins immediately after birth. These are
apocrine-type glands which are fully functional in infancy,
unlike the apocrine glands of the axilla and other skin
surfaces. This seems to suggest the existence of two distinct
types of apocrine glands that follow differing stages of
development.

5.4   Differences Between Standard Apocrine Glands and Ceruminous Apocrine Glands

The cartilaginous portion of the external auditory meatus
is lined by skin which contains two types of glands: the
sebaceous glands and the modified apocrine sweat glands
(ceruminous glands). Both types contribute to the formation of cerumen. Subcutaneous adipose fatty tissue is
abundant in the axillary region. In contrast, the skin of
the inner ear is thin, with scant fatty tissue and shallow,
superficial ceruminous apocrine glands. The apocrine glands
have a close relationship with the hair follicles. The ducts
of apocrine glands open close to the superficial ducts of the
sebaceous glands appended to the hair follicles. The ducts
of the ceruminous glands, however, are independent of
the sebaceous glands and open directly, close to the skin
surface (Fig. 5.1).

The ceruminous glands are larger when the cerumen is
wet, the lumen of the gland is wider, and the epithelial
layer is thicker. The ceruminous glands are lined by simple
cubical epithelium cells with round nuclei. Histologically,
the ceruminous gland is a major sweat gland comparable to
the axillary apocrine gland. The morphological structure of
ceruminous glands is quite similar to that of the axillary
glands but very different from eccrine sweat glands.

Many researchers, especially Nagashima (1934), have
reported a close relationship between bromidrosis odor
and the fatty content of ceruminous gland wax. Abundant
lipid droplets and pigment granules in the cytoplasm of the
secretory cells are found in those individuals who have
wet cerumen, but are scant in those with dry cerumen.
Secretory cells which have striated cuticular borders are
characteristic in cases of wet cerumen, but in cases of
dry cerumen these striations are found only in a very few
secretory cells.

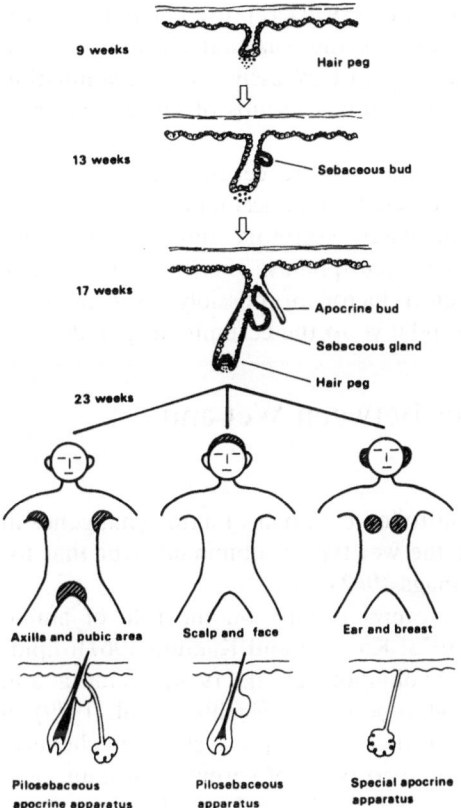

Fig. 5.1. Pilosebaceous and apocrine glands

In histochemical findings, the chemical characteristics of the fatty content are the same in apocrine glands and in ceruminous apocrine glands. Fat quantity is significantly higher in bromidrosis cases. Iron reaction is more evident in cases of thicker epithelium and excessive gland secretion. Succinate dehydrogenase activity can be observed in normal ceruminous glands as a faint positive granular reaction in their lining cells. Enzyme activity is considerably increased in cases of excessive wax accumulation. The normal glands show a moderate reaction for alkaline phosphatase in the apical cell area lining the glands. In cases of excessive wax, however, the reaction is very strong in that cell area, and the remainder of the cell is full of wax granules. Furthermore, while normal glands show moderate alpha esterase activity in the form of fine granules within the lining cells, this enzyme activity is increased in cases of excessive wax, as evident in the intense positive reaction which fills the cells.

Mandour and associates (1974) reported that the histological structure as well as the enzymatic activity of the ceruminous glands in cases of excessive wax accumulation do not differ in normal control groups of either sex or of different ages.

The enzymatic activities of the ceruminous glands are increased in cases of excessive wax accumulation, indicating hyperactivity and increased secretory function. The observed increase in acid phosphatase enzyme activity may point to an etiological factor of possibly exogenous or indigenous irritation relative to the ceruminous glands.

5.5   Differences Between Wet and Dry Cerumen

This difference is controlled by a pair of autosomal genes in which the allele for the wet type is dominant over that for the dry type (Matsunaga 1962).

There have been several reports on the role of amino acids (Bauer et al. 1953; Kataura and Kataura 1967), lipids (Nagashima 1934; Akobjanoff et al. 1954; Kataura and Kataura 1967) and glycopeptides (Shichijo et al. 1979) in human cerumen, but there is still no clear biochemical basis to distinguish the two types of cerumen. Kataura and Kataura (1967) analyzed cholesterol, triglycerides, and free fatty acid in the cerumen of each type, and reported no qualitative or quantitative differences in these lipids between the two types.

Suzuki et al. (1985) found a substance with a positive reaction to a resorcinol reagent on a thin layer chromatogram. A molecular species of sialic acid was then isolated in a free type from wet-type cerumen but not from the dry type. These results imply that the compound in the wet cerumen is 2.7-anhydro-N-acetylneuraminic acid. This sialic acid was not detected in dry-type cerumen, so its formation in the wet type may be regulated by an autosomal dominant gene.

We investigated the differences of dry and wet cerumen in humans by analyzing the lipid composition of the two types using thin-layer chromatography (Inaba et al. 1987a). The multiple solvent system recommended by Downing (1968) was used in most of our analysis and produced good separation of cerumen lipid.

Figure 5.2.a indicates the appearance of a typical thin-layer plate after the resolved lipids have been charred. Squalene, sterol esters, wax esters, triglycerides, free fatty

Fig. 5.2.a–c. Differences between dry and wet cerumen analyzed by the lipid composition of each type. **a** Thin-layer chromatogram of lipid extracts of cerumen. *SQ*, squalene; *SE*, sterol esters; *WE*, wax esters; *UI*, unidentified; *TG*, triglycerides; *FA*, fatty acids; *CH*, cholesterol, **b** Thin-layer chromatogram of nonpolar lipids of cerumen. *SQ*, squalene; *SE*, sterol esters; *WE*, wax esters; *UI*, unidentified; *WD*, wax diester, **c** Results of the thin-layer chromatogram

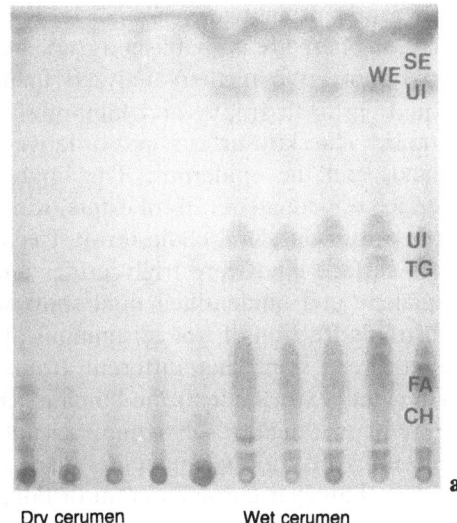

Dry cerumen   Wet cerumen

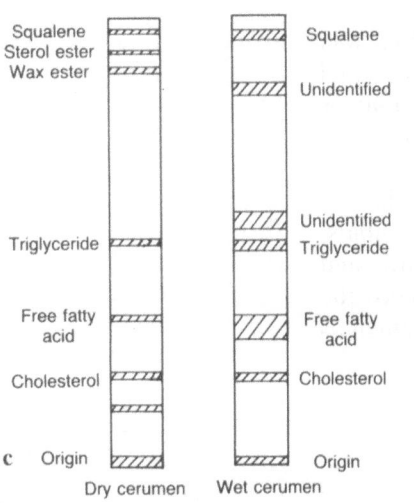

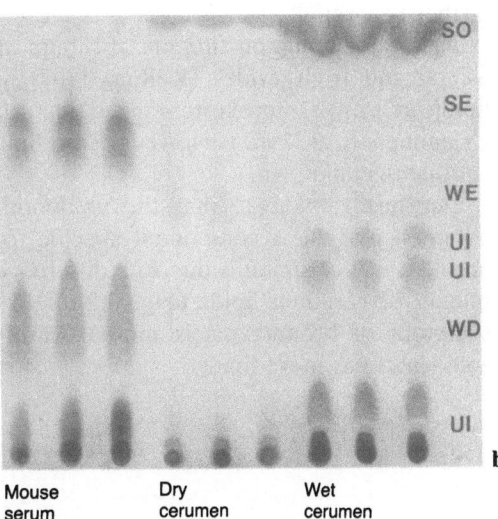

Mouse serum   Dry cerumen   Wet cerumen

acids, and cholesterol were found in the dry type. However, squalene, triglycerides, free fatty acids and cholesterol formed the main demonstrable fractions in the wet type. In wet cerumen, sterol esters and wax esters were not found in as great a quantity as compared with dry cerumen. In addition, two unidentified lipids were always present in wet cerumen. Because the two unidentified lipids were less polar than triglycerides, lipid extracts were fractionated by Sep-Pak silica cartridge. Non-polar lipid evaluates were analyzed by the method of Nikkari and Valavaara (1970).

However, the unidentified spots in wet cerumen have

different patterns with wax diesters and were further separated. There are no published studies on these unidentified lipid spots. We further analyzed them by means of gas-liquid chromatography and high-pressure liquid chromatography. The skin surface lipid is derived from the sebaceous glands and the epidermis. The lipid composition of skin surface was squalene, sterol esters, wax esters, triglycerides, free fatty acids, and cholesterol. The main constituents of skin surface lipid were triglycerides and wax esters. While squalene and unidentified lipid spots were the main demonstrable fraction of wet ceruminous lipids, the ceruminous lipids were somewhat different from skin surface lipids. The lipid extractable from human cerumen with organic solvents is a mixture of sebum, cerumen, and of lipid produced by the keratinizing epidermis.

The thin-layer chromatogram of nonpolar lipids is shown in Fig. 5.2.b. Wax diester was the major lipid in mouse sebum. However, the unidentified spots in wet cerumen have a pattern different from that of wax diester and are farther separated.

The major lipid constituents of sebum are squalene, wax esters, and triglycerides (Kellum 1967) and in epidermal lipids are sterols, sterol esters, glycerides, and phospholipids (Nieminen et al. 1967). However, the lipid composition of cerumen is unknown.

Our results indicate that the unidentified lipids in wet cerumen may be a component specific to cerumen lipids, and that wet cerumen is due to a difference of quantity and quality of cerumen lipids (Fig. 5.2.c). In cases where the symptom of bromidrosis is more severe, the cerumen is softer and has more lipids.

Chapter 6. Differences Between Bromidrosis and Nonbromidrosis Patients

6.1 Lipid Analysis of Axillary Skin Surface Lipid of Patients with Bromidrosis in Japan and Korea

The recent development of lipidic analysis has led researchers to investigate the formation and composition of dermal surface lipid. Therefore, many reports have described the correlation between the lipid formation in sebaceous glands and that on the surface of skin in various dermatoses. However, research and reporting on the dermal surface lipid of patients with bromidrosis has been relatively uncommon.

Bromidrosis is caused mainly by apocrine gland secretion. Also related to its etiology are eccrine sweat, sebaceous secretion, and bacteria. The lipid etiology in the axillary fossa has not yet been completely elucidated.

Choi and Inaba (1984b) made an analysis of normal men and bromidrosis patients (Japanese and Korean) in order to observe how skin surface lipid affects the occurrence of offensive axillary odor. The result of the lipid analysis of the ten normal Japanese men was triglycerides 26.8% ± 4.4% and cholesterol 5.4% ± 2.0%, and that of the bromidrosis patients was triglycerides 25.6% ± 4.0% and cholesterol 5.7% ± 2.1% (Table 6.1.a). No difference was observed in the levels of both triglycerides and cholesterol between the normal men and the bromidrosis patients; additionally, no phospholipids were observed. No significant difference was observed among the Koreans, as indicated in Table 6.1.b.

We also compared the levels of triglycerides, cholesterol, and phospholipids with the data set forth by Downing et al. (1969) and Green et al. (1970) but, again, no significant

Table 6.1.a,b. An analysis of the effect of skin surface lipids on the occurrence of offensive axillary odor in **a** normal and bromidrosis Japanese males and **b** normal and bromidrosis Korean males. (Reproduced with permission from Choi et al. 1984b)

a

Lipids	Control (n = 10)	Bromidrosis (n = 20)
Triglycerides	26.8 ± 4.4	25.6 ± 4.0
Cholesterol	5.4 ± 2.0	5.7 ± 2.1
Phospholipids	0	0

b

Lipids	Control (n = 10)	Bromidrosis (n = 20)
Triglycerides	26.4 ± 5.0	24.4 ± 6.0
Cholesterol	5.7 ± 1.9	6.3 ± 3.4
Phospholipids	0	0

Table 6.3. No difference was found between cases of bromidrosis with odor and those without odor in distribution of triacylglycerols of sebaceous glands

Classes of triacylglycerol (by fatty acid)	Bromidrosis with odor (%)	Bromidrosis without odor (%)
MMA, LLA, MLA	5.4 ± 1.3	5.9 ± 1.5
LLL	18.5 ± 1.9	20.6 ± 1.8
OLL	29.5 ± 1.2	30.9 ± 1.9
PLL	31.0 ± 1.9	32.1 ± 1.8
OOL, PPL, POL	14.1 ± 0.2	14.5 ± 0.4

Triacylglycerol fatty acids: M, myristic; L, linoleic acid; O, oleic acid; P, palmitic acid; A, arachidonic acid. (Reproduced with permission from Choi et al. 1984a)

difference was observed (Table 6.2). It may be presumed that the absence of phospholipids indicates that they have been converted to fatty acids.

The above findings indicate that sebum, like the difference in the apocrine glands themselves, does not play any significant role in inducing underarm odor.

6.2 Determination of Triacylglycerol in Axillary Sebaceous Glands of Bromidrosis Patients

Choi et al. attempted to analyze triglycerides as the main component of sebaceous lipid between nonbromidrosis and bromidrosis patients (Choi et al. 1984a). In order to obtain the pure lipid, sebaceous glands were protruded and re-

Table 6.2. Comparison between the Downing, Green, and Choi reports on the composition of triglycerides and cholesterol. No difference was observed.

	Downing (1969)	Green (1970)		Choi et al. (1984b)	
		Upper limbs	Lower limbs	Axilla of Japanese bromidrosis patients	Axilla of Korean bromidrosis patients
Triglycerides	19.5 ~ 49.4	30.7	24.2	26.8 ± 4.42	25.6 ± 4.00
Diglycerides	2.3 ~ 4.3	1.5	1.8		
Fatty acids	7.9 ~ 39.0	36.4	37.8		
Squalene	10.1 ~ 13.9	6.9	6.2		
Wax-esters	22.6 ~ 29.5	15.8	12.9		
Cholesterol	1.2 ~ 2.3	4.1	9.4	5.4 ± 1.95	5.7 ± 2.06
Cholesterol-esters	1.5 ~ 2.6	4.4	7.5		
Phospholipids	—	—	—	—	—

(Reproduced with permission from Downing et al. 1969, Green et al. 1970, and Choi et al. 1984b)

moved by a subcutaneous tissue shaving procedure from bromidrosis patients with wet cerumen and from hyperhidrosis patients with dry cerumen. After isolating the individual sebaceous gland by the Kellum method (1966), lipid extraction was carried out by the Folck method.

This extraction was analyzed by high-pressure liquid chromatography and gas-liquid chromatography. Distribution of triacylglycerols of sebaceous glands in cases of bromidrosis with or without odor, as shown in Table 6.3, includes dimyristoarachidonin (MMA), dilinoleioarac-

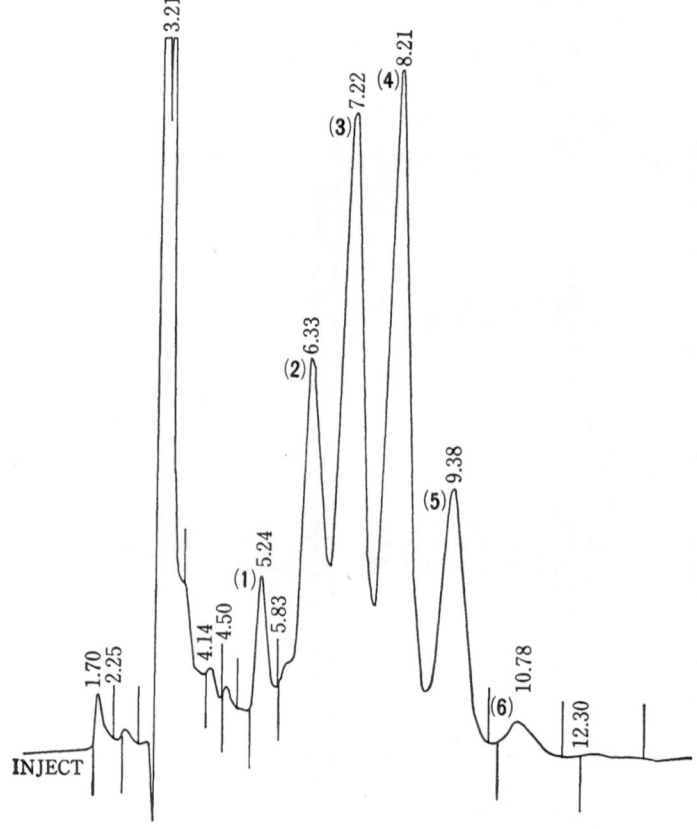

Fig. 6.1. Analysis of lipids in sebaceous gland of control subject by high-pressure liquid chromatography. Triglyceride fatty acids: *M*, myristic; *L*, linoleic; *O*, oleic; *P*, palmitic; *A*, arachidonic. Peak (1); triglyceride (MMA, LLA, and/or MLA), peak (2); triglyceride (LLL), peak (3); triglyceride (OLL, and/or an unknown triglyceride), peak (4); triglyceride (PLL), peak (5); triglyceride (OOL, PPL, and/or POL), peak (6); unidentified component, not a triglyceride. Reproduced with permission from Choi et al. 1984a

hidonin (LLA), and myristolinoleioarachidonin (MLA) complex. No difference was found between cases of bromidrosis with odor (5.4% ± 1.3%) and those without odor (5.9% ± 1.5%). In addition, there were no differences in the classes of trilinolein (LLL), dilinoleioolein (OLL), dilinoleiopalmitin (PLL) and the complex of dioleiolinolein (OOL), dipalmitolinolein (PPL), and almitooleiolinolein (PDL).

High-pressure liquid chromatography analysis of lipids in the sebaceous glands of normal controls is shown in Fig. 6.1. These findings indicate that five classes of triacylglycerol fractions were separated and that the chromatographic patterns of both groups turned out to be identical in nature.

6.3   Differences Between Bromidrosis-Type and Nonbromidrosis-Type Apocrine Glands

Bromidrosis-type apocrine glands are larger in number, have higher cell walls, smaller gland apertures, and obvious secretory capability (Fig. 6.2.a). On the other hand, nonbromidrosis-type apocrine glands tend to be relatively few in number (see Fig. 6.4.a) (Harada and Inaba 1985). They are characteristically flattened, expanded in diameter, and diastolic, with secretory cell walls low in height and forming a large secretory canal. Electron microscopy reveals that the capability of secretory function remains in the cells (Fig. 6.4.a–c). It is not yet known whether the causative mechanism of bromidrosis is based on the density of gland numbers or on the quality of the gland secretory function.

6.3.1   Bromidrosis-Type Apocrine Glands

Thin-Sliced Specimen. There is a single layer of columnar or cubic glandular cells in a line, the myoepithelial cells are distributed intermittently, and the nucleus is observed sporadically (Fig. 6.2.a,b).

Electron Microscopic View. The nucleus of the glandular cell is ball-shaped and distributed over the basal surface (Fig. 6.3.a). In the relatively large portion of cytoplasm in the upper margin of the nucleus, dark granules, rather irregular in shape, or a lysosome-like body can be observed. These are secretory granules, and tend to increase in proportion to the numbers of nonbromidrosis-type apocrine glands. Also, a large Golgi apparatus is detected in the

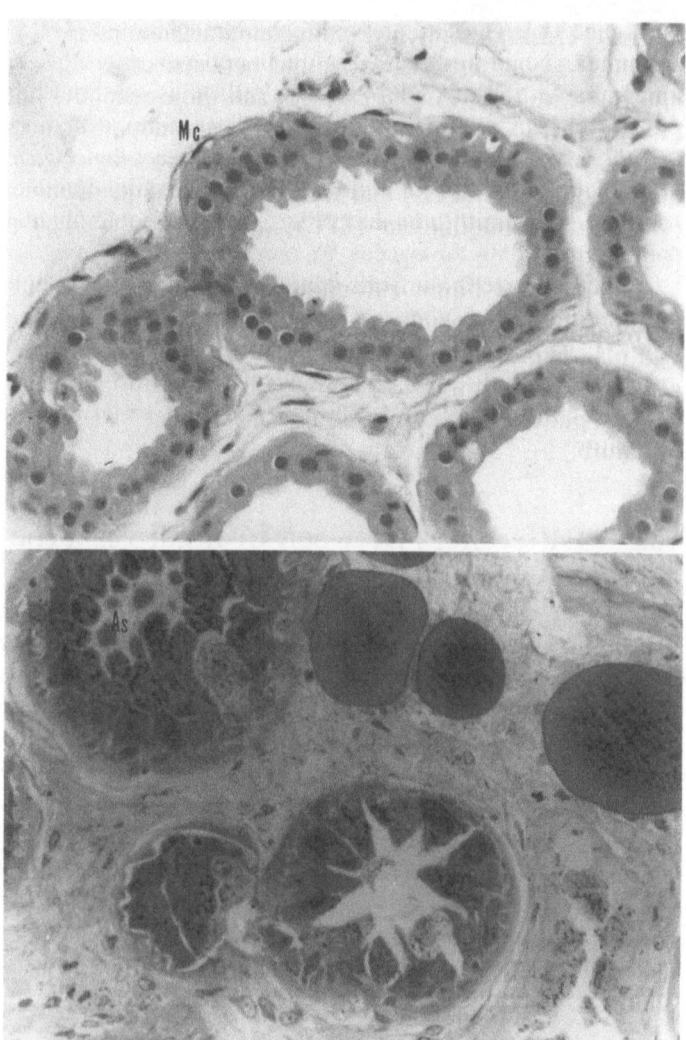

Fig. 6.2.a,b. Thin-sliced specimens of the bromidrosis-type apocrine glands. **a** Photomicrograph of a human axillary sweat
gland. *Mc*, myoepithelial cells, **b** 1 μm specimen stained by tolui-
dine blue. The glandular cortex cell at the secretory portion is
larger and higher in size and the lumen is narrower in comparison
with the nonbromidrosis-type specimen

Fig. 6.3.a,b. Electron micrograph of the secretory portion of the
bromidrosis-type apocrine glands. **a** Cylindrical cells contain the
nucleus of the glandular cell (*Nu*), large dense granules (*Gr*),
myoepithelial cells (*Me*), and infolding of the basal plasma membrane (*IF*) as seen at the bottom of the figure. *Lu*, lumen; *Bm*,
basal membrane; *Co*, connective tissue; *Bac*, bacillus, **b** Enlarge-

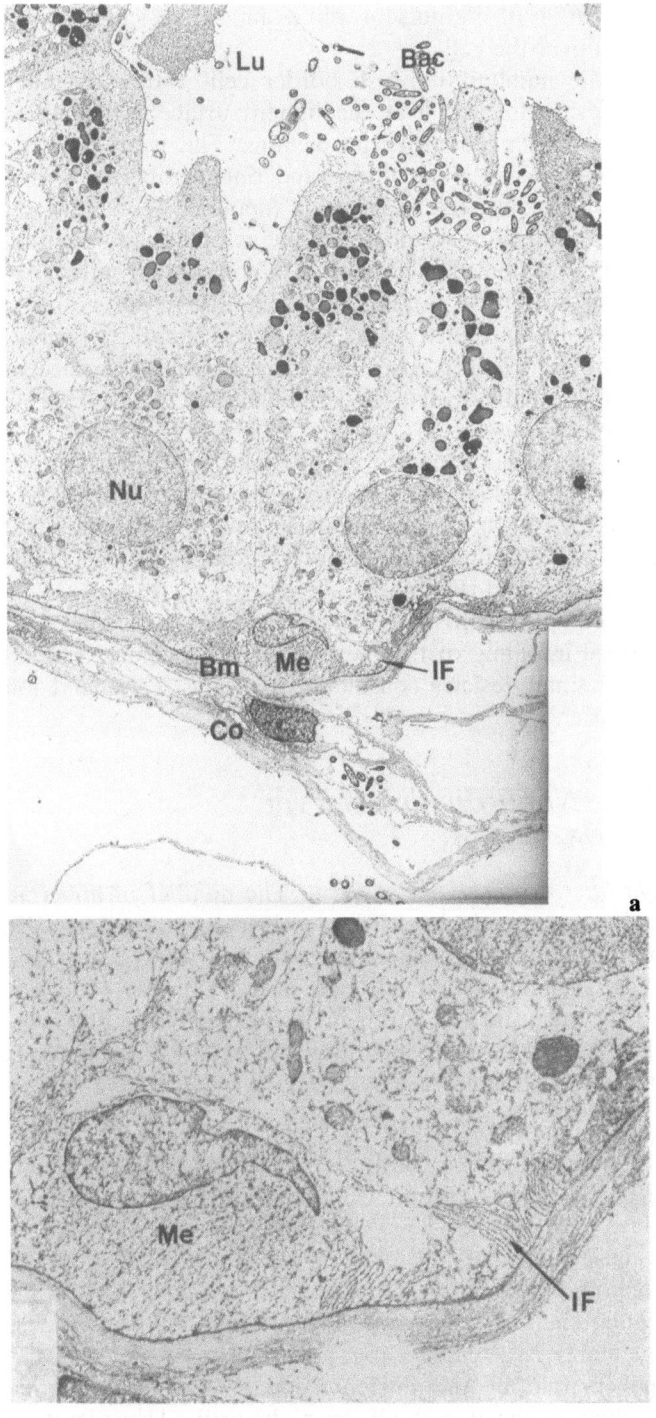

ment of basal infolding (*IF*). Basal parts of the secretory cell with
complicated foldings of the plasma membrane. Me, myoepithelial
cell. (Reproduced from Harada J, Inaba M 1985)

upper margin of the nucleus and a smooth vesicle is developed all over the cell.

A large number of brush border cells rather irregular in shape are formed over the lumen surface, which form tongue-like or balloon-like projections. This is an apocrine projection. When part of the apocrine cytoplasm is discarded, it is converted to secretion (apocrine secretion).

The frontal area of the cell body cytoplasm dominating the lower portion of microvilli does not contain mitochondria and secretory granules, and can be observed as a homogenous substance under low magnification, which is, therefore, called the crust. The external border surface of the cell is relatively flat, and intercellular space is rarely observed.

Most of the basal surface of glandular cells is contiguous to the myoepithelial cell, although some part of it is directly attached to the basement membrane at the myoepithelial gap. Here, a well-developed basal infolding plasma membrane is observed (Fig. 6.3.b).

According to Kurosumi et al. (1959), the front portion of the basal infolding of the membrane has become wrinkled lines of small vesicles considered to act as water and ion transporters.

6.3.2   Nonbromidrosis-Type Apocrine Glands

Light Microscopic Point of View. The number of apocrine glands is extremely low. The secretory portion is usually loosened (Fig. 6.4.a). Similar to bromidrosis-type apocrine glands, columns or cubic single-layer cells are lined up. Most of them are low in height and appear flat; however, cells of greater height are also observed. The lumen is loose and enlarged (Fig. 6.4.b).

Electron Microscopic Point of View. Similar to the bromidrosis-type apocrine gland, the nucleus of the glandular cell is ball-shaped and unevenly distributed over the basal surface (Fig. 6.4.c). The cytoplasm at the upper portion of the nucleus is relatively small in amount compared to the cytoplasm of the bromidrosis type. Dark granules of rather irregular shape and lysosome-like granules are observed in large numbers. A large Golgi apparatus is observed in the upper margin of the nucleus, and smooth endoplasmic reticulum is developed all over the cell. The secretory granules are small in number and apocrine secretion seems to occur.

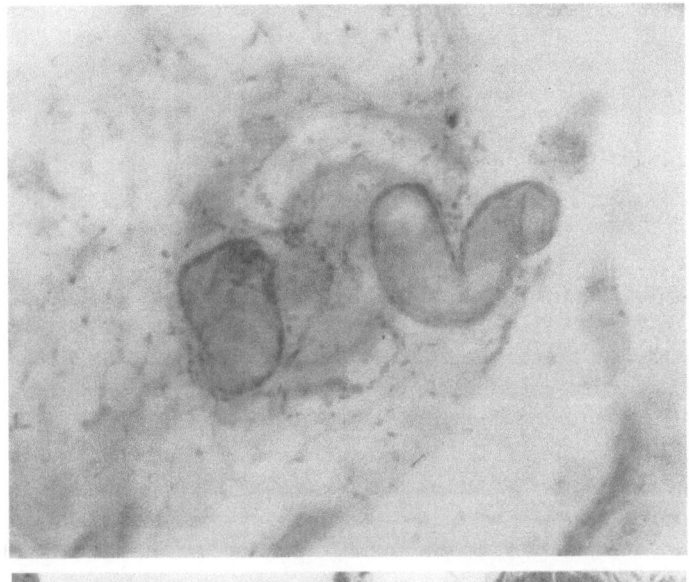

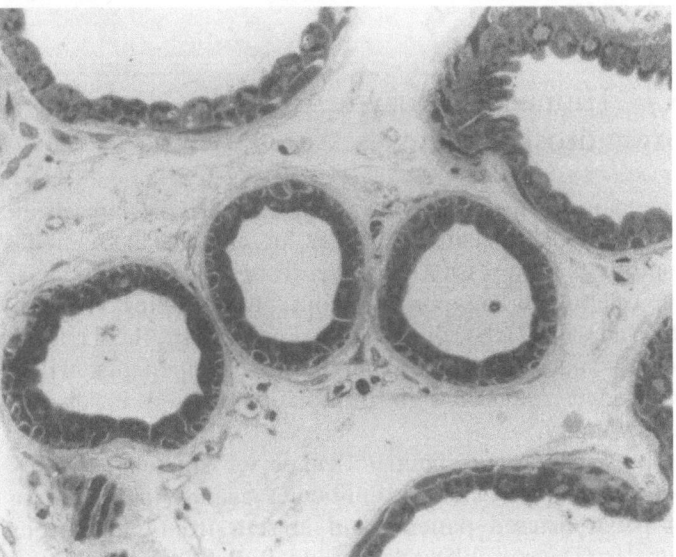

Fig. 6.4.a–c. Photomicrographs of human axillary sweat glands (nonbromidrosis-type apocrine glands). **a** The number of apocrine glands is extremely low, **b** The secretory portion is usually loosened and enlarged. Columns of cubic single-layer cells are lined up, **c** Electron micrograph of the secretory portion of nonbromidrosis-type apocrine gland. *Gr*, secretory granules; *G*, Goldi apparatus; *Me*, myoepithelial cell; *Nu*, nucleus. (*For Fig. 6.4.c see following page*)

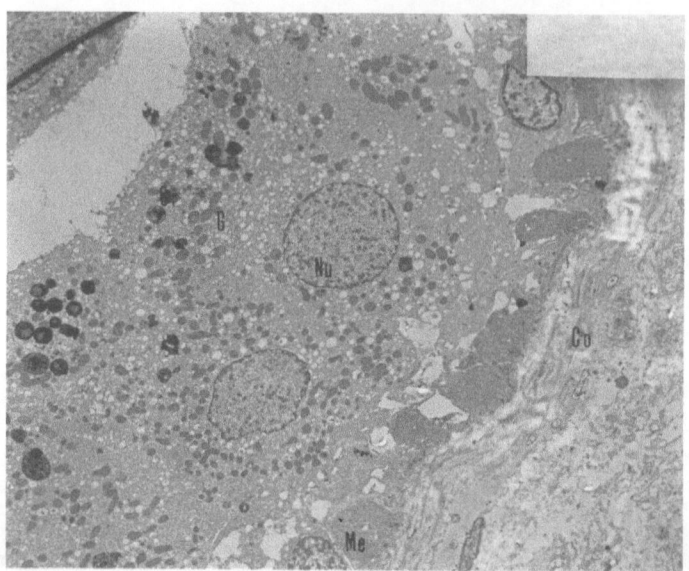

Fig. 6.4.c

6.4 Human Leukocyte Antigens (HLA) in Bromidrosis Patients

The major histocompatibility complex (MHC) is present in all species which exhibit cellular immunity. This complex controls the expression of three groups of gene products: (1) cell membrane polymorphous glycoproteins HLA-A, -B, and -C; (2) B-cell antigens produced by the D locus genes, and (3) several complement components. HLA is strongly implicated in transplantation and in disease susceptibility.

Classification of human leukocyte antigens initially depended upon detecting antibodies against them in sera of polytransfused patients and in sera from multiparous females (Dausset 1958; Payne and Rolfs 1958; Van Rood et al. 1958). Serological classification and detection of HLA specificities were further accelerated by the establishment of World Health Organization International Workshops (Bodmer et al. 1975). Monoclonal antibodies produced in the hybridoma system (Barnstable et al. 1978) may provide even more specific typing reagents (Duvic and Goldsmith 1983).

Inaba and associates (Inaba et al. 1985a) conducted a joint study of the hereditary factor in bromidrosis, especially in relation to the HLA antigen, and the distribution of HLA-A and -B antigens in wet cerumen of bromidrosis patients was compared with that of normal Japanese subjects selected at random.

HLA-A and -B antigens were typed and all lymphocytes were tested using Terasaki typing trays and the standard NIH two-stage lymphocytotoxicity technique (Terasaki and McCelland 1964) for a series of 34 patients. The results of the HLA antigen and gene frequency studies of 34 Japanese bromidrosis cases are shown in Table 6.4 while Table 6.5 covers 91 controls in a prior study conducted in the Kanto district (Tsuji et al. 1979).

Only a slight difference was observed between the bromidrosis and the control groups at the HLA-A locus, excluding those in A9, Aw24 and A26. The gene frequency of A9 for bromidrosis was 0.3583 (Table 6.4), whereas that for the control group was zero. However, the gene frequency of Aw24 for the control group was 0.3124 (Table 6.5).

The A9 for bromidrosis cases is either A9 or Aw24 for the Kanto controls, and therefore no statistical difference could be found. The gene frequency of A26 for the controls was 0.0899, whereas no A26 was typed for the bromidrosis cases. This absence of A26 in the bromidrosis group can be attributed to a different typing criterion between the two laboratories. By pooling A10 and A26, the discrepancy between the two groups in A10 and A26 can be deleted. For the bromidrosis cases and the Kanto controls at the B locus, the B5 and B40 gene frequencies in the bromidrosis cases were found to be much higher than those in the Kanto control group. They were 0.1598 versus 0.0055 in the B5 and 0.1775 versus 0.0330 in the B4 frequency. However, if the B5, Bw51, and Bw52 frequencies are pooled, there

Table 6.4. HLA antigen and gene frequencies of 34 bromidrosis cases in Japan. (Reproduced with permission from Inaba et al. 1985a)

HLA-A	Percent of Antigen Frequency	Gene Frequency	HLA-B	Percent of Antigen Frequency	Gene Frequency
A2	35.79	0.1956	B5	29.41	0.1598
A9	58.82	0.3583	Bw51	2.94	0.0148
A10	17.65	0.0925	Bw52	2.94	0.0148
A11	26.47	0.1425	B7	17.65	0.0925
Aw31	5.88	0.0299	B12	17.65	0.0925
Aw33	12.55	0.1255	Bw44	5.88	0.0298
Blank	—	—	B15	17.65	0.0925
			Bw22	17.65	0.0925
			Bw35	14.71	0.0765
			B40	32.35	0.1775
			Bw48	2.94	0.0148
			Blank	—	—

Table 6.5. HLA antigen and gene frequencies of 91 Japanese in the Kanto District of Japan. (Reproduced with permission from Tsuji et al. 1979)

HLA-A	Gene frequency	HLA-B	Gene frequency
A1	0.0055	B5	0.0055
A2	0.2334	Bw51	0.0876
A3	0.0055	Bw52	0.0688
A9	—	Bw53	0.0055
Aw23	—	B7	0.0495
Aw24	0.3124	B8	—
A10	0.0114	B12	0.0745
A25	—	Bw44	0.0330
A26	0.0899	Bw45	—
A11	0.0954	B13	0.0055
A28	—	B15	0.1161
Aw19	0.0114	Bw16	—
Aw30	—	Bw38	—
Aw31	0.0508	Bw39	—
Aw33	0.0975	Bw39.1	0.0220
Aw34	0.0055	Bw39.2	—
Blank	0.0814	B17	0.0226
		Bw22	0.0220
		Bw54	0.0518
		B27	—
		Bw35	0.0971
		B37	0.0055
		B40	0.0330
		Bw48	0.0110
		Bw42	—
		Bw46	0.0394
		Bw40.1	0.0646
		Bw40.2	0.0778
		B8 J	—
		Blank	0.0963

would be no significant difference between the two groups: 0.1894 for bromidrosis cases versus 0.1619 for controls. The B40 used in typing bromidrosis samples was evidently subdivided into B40, Bw40.1 and Bw40.2 when the controls were typed.

Any difference between the bromidrosis and the control groups may be due to a difference between the two laboratories in typing procedure and/or new designations of HLA alleles over the years rather than any real difference between them in the HLA system. It can be concluded that there was no strong evidence for HLA association with bromidrosis when the Japanese bromidrosis data was compared with the published Kanto control data.

Chapter 7. Relationship Between Bromidrosis and Sense of Smell

Human olfactory sensation is one of the most acute sensory perceptions and can even detect a 0.002 ppm dilution of methylmercaptan which is the source of the offensive feces odor (Yamamura 1981).

There is a close interaction between the sense of smell and bromidrosis. Most bromidrosis patients are concerned about their own body odor because they can detect it themselves. But disorders in odor perception sometimes exaggerate the condition. Patients who have an especially keen sense of smell (hyperosmia) can become overly concerned about their own body odor. Patients with hyposmia, on the other hand, may not perceive even strong body odor and may realize they have it only when other people tell them.

Smell is important to all mammals. Its manifold role includes recognizing food, locating food sources, recognizing direction and location by spoor detection, confirming one's own species, detecting enemies, and pheremones and sexual identification. Among pheromones, in particular, the best known is the sex hormone emitted by female moths to attract males. Male moths often travel long distances in search of females, attracted by a small number of pheromone molecules drifting in the air. Scent glands give many animals their unique smell and are either apocrine or apocrine-sebaceous glands similar to those of the human axilla (Table 7.1). The prominent social functions of these glands form a complex system of chemical messages (pheromones) that stimulate particular types of behavior. The glands are used to delineate territory, express dominance, produce sexual excitement in the male or sexual attraction in the female, repel enemies, and also serve as individual signatures.

Not only the olfactory sensation but the visual and audi-

Table 7.1. Mammalian scent glands

Mammal	Gland (apocrine-sebaceous)	Odorant	Reference
Rabbit "odor"	Anal/apocrine	Cis-undec-4-enal	Goodrich et al. 1978
Beaver	Castor	Castoramine	Valenta and Khaleque 1959
Elephant	Temporal/apocrine		Adams et al. 1978
	Interdigital	Short-chain-acids	Anderson et al. 1978, Brundin et al. 1978
Deer	Subauricular	Isovaleric acid	Müller-Schwarze et al. 1976
	Tarsal	Aldehydes	Anderson et al. 1975
Musk deer	Sebaceous	Muscone	Do et al. 1975
		Bacterial action on secretion	
Human	Apocrine	16-3-Androstenone	Labows 1988
		Androstenol	Zeng et al. 1991
		ε-3-Methyl-hex-2-enoic acid	
Marmoset monkey	Circumgenital	Butyrate esters of long chain alcohols	Epple et al. 1979
Bactrian camel	Occipital	Isovaleric acid androst-16-en-3-one	Ayorinde et al. 1982

(Modified with permission from Labows 1982 and 1991)

tory senses are well developed in mammals, which means they do not need to depend completely on their sense of smell. However, it is also true that smell plays a vital role in determining the potency of sexual reproduction.

Male mice are sensitive to the vaginal secretion of female mice, and females prefer the smell of normal males to males that have been castrated. Male mice evidently discriminate their mates from their mother and sisters by olfactory sensation, and prefer females with no blood relationship.

Many mammalian species have the habit of marking within their own territory. Kimura et al. (1985, 1988) of Tokyo University reported that the laboratory mice had showed the habit of marking by urine. This habit was observed only among the males. Castrated males lose this marking habit; however, it is restored by androgenic injection, which indicates that it is under androgenic control. Androgen injection to females at an early stage postnatally will induce the acquisition of this marking habit in adulthood. It is conjectured that the action center of the two sexes for the marking habit is not determined at the time of birth, but depends on whether or not the brain receives an androgenic stimulus.

Male dogs have the habit of marking by urine. This can be observed exclusively in males, and not in any normal adult females. A study of the androgenic effect on the fetus of a pregnant female dog was studied. As a result of the androgenic injection, the female cub, when it reached adolescence, had acquired the habit of urine marking by lifting its hind leg. This indicates that the acquisition of this habit is determined by whether the brain, which controls the marking habit of dogs, comes under the influence of androgen during its development. The fact that males have the marking habit while females do not indicates that the males are born with an instinct that has a tendency to establish territories.

Odors are basically classified as pleasant or unpleasant. This is because the olfactory sensation, after passing through the nostrils, is perceived and its data processed in the cerebrum's limbic system which controls emotional and instinctive activities. A more acute sense of discrimination occurs such as when selecting wines with the greatest care. The data processed in the cerebrum's limbic system is transferred to the cerebral cortex for further processing. However, emotion generally precedes reason in the olfactory sensation. Intense odors are offensive to olfactory sensation while aromatic odors that are perceived as pleasant and sweet to the sense also have appropriate intensities.

Threshold is applied as the index. The thresholds for such powerful smells as skatole and mercaptan are extremely low (Takagi 1988). With respect to human beings, however, this difference can be strongly influenced by upbringing. Moreover, humans walk upright and have a diminishing sense of smell. In terms of food, however, a substantial loss of smell could be a great disadvantage as they have a very important role in human life. Food spoilage could be difficult to detect without a sense of smell and, if ingested, could be poisonous.

In addition, human beings have emotional complexities: a "good smell" can induce a good mood. Odor evaluation can be strongly influenced by one's environment. Among some people, the odor of newly turned soil seems to be invigorating. In Japan, however, it is considered unpleasant. The Japanese dislike body odor as well, but there are other people who do not necessarily consider the odor offensive.

In anosmia (the loss of the sense of smell), the patient may not be aware of his condition, mistaking olfactory anesthesia for loss of taste (ageusia). That should be understandable, since the sense of taste depends a great deal on volatile substances contained in foods and beverages. The sensation of flavor is, in fact, an amalgam of smell and taste. A similar effect can occur in cases of upper respiratory infection, due to clogged nasal passages or damaged olfactory mucosa and nerve filaments.

The condition of parosmia, which perverts the sense of smell, may occur in local nasal complications such as paranasal sinus empyema. However, this phenomenon is not invariably abnormal since unpleasant odors may be persistent and reoccur because of other olfactory stimuli.

Olfactory hallucinations can be traced to a central organic origin. An aura that occurs in an epileptic seizure may be perceived as an offensive odor. In these cases, the lesion is most often located in the inferior and medial zones of one of the temporal lobes. Schizophrenic patients may insist they have offensive body odors, but these claims are almost always delusions.

In testing for odor perception, the patient is instructed to close his eyes. One nostril is blocked and then the other while he sniffs several test odors. The olfactory nerves are not impaired if the patient can detect these odors in each nostril, even though he may not be able to identify them. Ammonia should be omitted from odor perception tests because the pungent odor it emits tends to stimulate the trigeminal nerves even in cases of complete anosmia.

7.1   Sense of Smell Activation

When and as substances stimulate the nostril mucosa, sensitized tissue cells detect odor and convert it to electrical energy. Transmitted through thousands of cells, this energy arrives in the brain and is received as a chemical stimulant. The mind decides whether the odor thus perceived is pleasant or unpleasant. Why odors stimulate those sensitized nasal cells has not been elucidated, but there have been some attempts at explanation.

One of them is related to absorbency. If, for instance, an odor particle is small enough to be soluble in water, it can be absorbed by nerves and fatty tissues. Tissue which is odor sensitive then senses the chemical result the fatty tissue has produced. This cycle extending from air-borne odor to chemical agent in fatty tissue is, however, riddled with problems that have not been solved. Another supposition points to a pigment and particle reaction which produces vibratory waves. The waves in turn stimulate tissues. A third explanation, called the chemical theory, makes note of the fact that just as there are three primary colors and the sense of taste can be classified into four primary taste types, a case could also be made for a primary classification of odors.

7.2   Offensive Bromidrosis Odor

Human beings differ in their odor perception and evaluation, but ordinarily respond favorably to odors such as flowers, musk, peppermint, and so forth. On the other hand, the odors of carbon bisulfiteformalin and ethyl mercaptan (C_2H_5SH, a volatile colorless, transparent liquid having an offensive garlic-like odor) are almost always considered unpleasant. These olfactory evaluations are the same for both women and men.

Bromidrosis odor is generally disliked. In a survey of what the bromidrosis patients thought of the odor, the result was that 90.1% of the patients disliked it. Among nonbromidrosis controls, 92% responded that they disliked the odor as well (Fig. 8.16). The perceived strength of odor can be an important factor in evaluation. An unpleasant odor can be too weak to be offensive for more than a few moments. On the other hand, a pleasant odor which is overly strong, such as too much perfume, can also be offensive. Relative odor strength and olfactory sensitivity are other factors which contribute to classifying an odor as pleasant or unpleasant. A slight body odor may seem erotic to the opposite sex.

Many patients who claim that they have strong body odor tend to exaggerate the condition. Even patients who have received surgical treatment for bromidrosis (for example, the complete excision method) may still claim they have detectable odor, but the tiny amount of odor that remains does not bother anyone else.

7.3   Odor Measurement

The question, "Are you sensitive to your body odor?" was posed to bromidrosis patients: 67.7% of them said that they were. The rate for nonbromidrosis patients who replied affirmatively was 81%. The lesser rate for bromidrosis patients may be explained in terms of their being accustomed to body odor. A public restroom, for example, may first smell repugnant, but if we stay inside it long enough, we may become so accustomed to the odor that we may no longer notice it. Bromidrosis patients may be less sensitive to body odor because being accustomed to it, they have lapsed into selective olfactory fatigue.

There is no universal standard as yet to measure odor with the same precision found in universal standards for measuring vision and audition. However, studies of the chemistry of odors and the physiology of odor production are expanding (Amoore 1977; Labows 1979). Using gas chromatographic and spectrometric methods, researchers are on the verge of identifying the chemical structures of some odoriferous substances. Odor detection in medical diagnosis still depends on the physician's olfactory sensibility. There are two types of odor sensitivity: the dilution method and the intravenous injection method. Strawberry extract or Alinamin (vitamin B_1 extracted from garlic) is used with the dilution method. The intravenous method measures the time that elapses from injection of B_1 extract to odor detection. Other substances have been employed, but odor differentiation can be difficult even when garlic is used.

In 1984 Doty et al. reported the development of a standardized "scratch 'n sniff" olfactory test called the University of Pennsylvania Smell Identification Test (UPSIT). With this self-administered test it is possible to quickly and accurately assess odorants without a need for complex equipment or chemical stores.

Doty found that women consistently rate the micro-fragrance samples to be more intense than men do. This is in agreement with many other olfactory studies, including those of breath odor (Doty et al. 1982) and axillary odor (Doty et al. 1978).

7.4 Standardized Olfactory Test Developed in Japan

Takagi (1974) has suggested a standard for odor classification. He cites ten distinct odors that are stable, unchangeable, and readily identifiable:

1. DL-Camphor (CAM)—refined anti-pruritic camphor that conveys a seemingly "cool" sensation
2. r-Undecanolactone (UND)—peach
3. Isovaleric acid (VAL)—odor of putridity
4. Cyclotene—burn odor to which human beings are instinctively sensitive
5. Skatole—defecation
6. β-Phenylethyl alcohol—rose
7. Exaltolide (oxacyclo hexadecan-2-one)—musk
8. Phenol—disinfectant
9. Diallyl sulfide—garlic
10. Acetic acid—vinegar

We used the first three of these odors in a clinical experiment. Absorbent paper was soaked with each of these three agents in one-tenth to one-thousandth dilution. Patients were examined on two points. Does the bromidrosis patient have the same degree of odor sensitivity as a nonbromidrosis patient? What degree of body odor engages the sense of smell?

The sampling was comprised of 154 bromidrosis patients (27 male, 127 female) with wet cerumen and 20 hyperhidrosis patients (6 male, 14 female) with dry cerumen. The result obtained with use of dl-camphor (CAM) was $10^{-3.03}$, with r-Undecanolactone (UND) $10^{-5.36}$, and with isovaleric acid (VAL) $10^{-6.01}$. Both CAM and VAL were detectable even at high levels of dilution.

The blind, as well as individuals specialized in odor determination, are especially odor sensitive. Blind controls scored 10^{-6} with CAM and specialists 10^{-8}. Bromidrosis patients turned out to have much the same degree of odor sensitivity as the nonbromidrosis patients, both groups scoring 10^{-3} on the whole (Fig. 7.1). There were no differences detected by age range or between males and females.

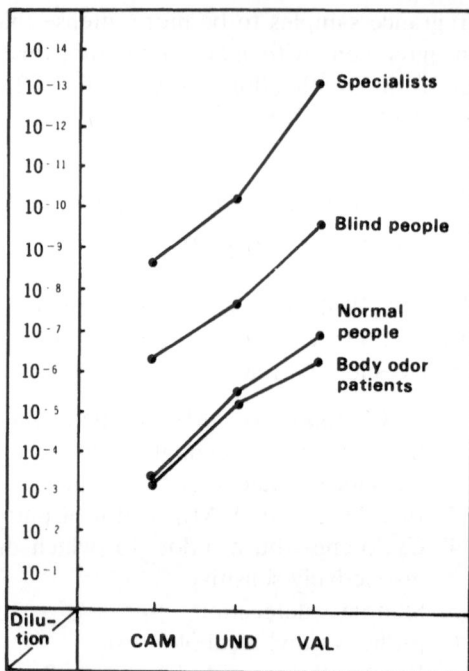

Fig. 7.1. Standardized olfactory test *CAM* (camphor), *UND* (peach odor), and *VAL* (odor of putridity) were used in testing body odor patients, normal people, blind people, and specialists for the degree of sense of smell. Body odor patients revealed little difference from normal controls in odor sensitivity. (Reproduced with permission from Inaba 1986)

In another experiment, nonbromidrosis patients were asked to smell the odor of bromidrosis patients. Those in the former group with average odor sensitivity did not detect the odor of bromidrosis. Alleged strength of body odor does have a psychosomatic aspect. Moreover, if odor sensitivity measures at 10^{-7}, that particular individual is far more odor sensitive than most people, and thus concludes that his own body odor is very strong. An especially high odor sensitivity can exaggerate the actual condition and engender neurosis in those who do have bromidrosis. As Fig. 7.1 shows, however, bromidrosis patients on the whole are much less odor sensitive than the blind or specialists.

Chapter 8.   Profiles of the Bromidrosis Patients in Japan

Because the per capita rate of bromidrosis is relatively low in Japan, those who are disposed to the condition tend to react with anxiety and seek permanent treatment for relief. The authors conducted a survey among patients who seek such treatment and uncovered a number of significant findings (Shimada et al. 1974, Inaba 1986).

8.1   Cases of Bromidrosis (Wet Cerumen) and Hyperhidrosis (Dry Cerumen) Classified by Age and Sex

Using the "subcutaneous shaving technique" we operated on 300 of the bromidrosis patients (238 females and 30 males), plus 30 female and 2 male hyperhidrosis patients. Most patients seeking permanent relief from bromidrosis and hyperhidrosis are between 19 and 25 years old and there are more female patients than there are male (Table 8.1). This does not necessarily indicate a higher frequency of these conditions in women, but may reflect the fact that women are more anxious to seek permanent treatment for relief. Even women over 60 years of age request the treatment because arthritis or similar afflictions make it difficult to suppress the odor by frequent bathing.

8.2   Relationship Between Physical Type and Bromidrosis

The authors classified patients by physical type as shown in Fig. 8.1. The results indicate that obese (30.3%) and

Table 8.1. Relationship between bromidrosis and hyperhydrosis according to age and sex

	Bromidrosis		*Hyperhidrosis*	
AGE	F	M	F	M
6~11 (Primary school)	4	0	0	0
12~14 (Middle school)	14	1	2	0
15~17 (Senior high school)	17	1	1	0
18~25	138	19	17	2
26~35	48	6	7	0
36~45	10	3	3	0
46~55	6	0	0	0
55~above 55	1	0	0	0
Total	238	30	30	2

(Reproduced with permission from Inaba 1986)

median (52.5%) physical types have a much higher incidence of bromidrosis than the very slender (14.1%) body types.

8.3   Recognition Source

In cases where both parents suffer from bromidrosis and the offspring are not aware of sharing that condition themselves, their friends are often reluctant to bring up the subject. Those who do learn of the condition from friends

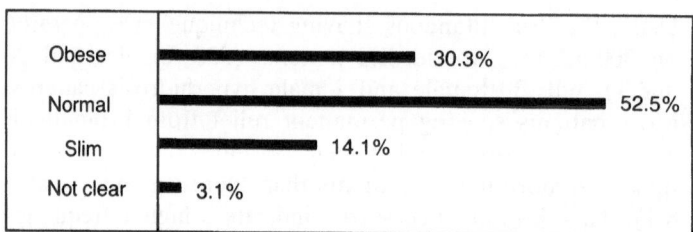

Fig. 8.1. Relationship between body constitution and bromidrosis. (Reproduced with permission from Inaba 1986)

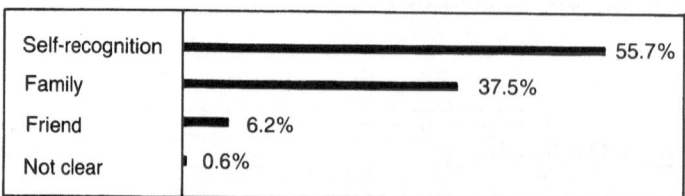

Fig. 8.2. Recognition of the condition. (Reproduced with permission from Inaba 1986)

experience acute distress at being singled out as different and, by inference, socially offensive. This frequently triggers enmity toward parents as the hereditary source of the condition (Fig. 8.2).

In cases of one-parent bromidrosis, approximately 50% of the offspring will develop the same condition. In those cases in which both parents have bromidrosis, the rate among their children is about 80%. Bromidrosis parents, particularly mothers, are anxious about passing it on. They will often detect the condition before the child becomes aware of it, and will then teach the child appropriate hygienic and cosmetic measures to deal with the problem. For instance, the odor characteristic of bromidrosis will diminish after taking a bath but reappear again within a few hours. The child is made aware of that time factor.

8.4 Relationship Between Bromidrosis and Seasonal Factors

As shown in Fig. 8.3, bromidrosis is influenced by seasonal changes. Bromidrosis patients are most aware of their condition at certain times of the year. Climate and temperature contribute to the levels of apocrine and eccrine secretions. When body temperature rises, the bromidrosis odor grows stronger. Bromidrosis patients are most aware of their condition in the hot summer months (74%). However, with the postwar development of Japanese living standards, indoor warmth is constant even in the winter season. This may explain the slightly higher recognition factor for winter (8.7%) as compared to autumn (6%). A psychosomatic factor can aggravate the condition. Patients concerned with concealing the condition from those around them in social situations often generate anxiety within themselves that contributes to a rise in eccrine and apocrine secretions, leading to a more pronounced odor.

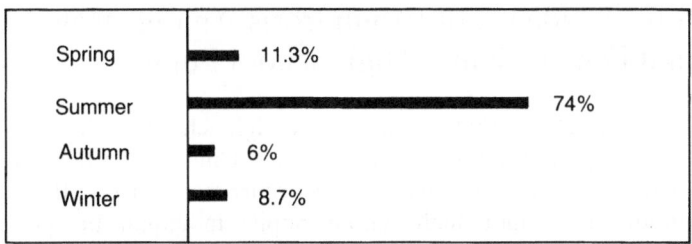

Fig. 8.3. Relationship between the seasons and bromidrosis. (Reproduced with permission from Inaba 1986)

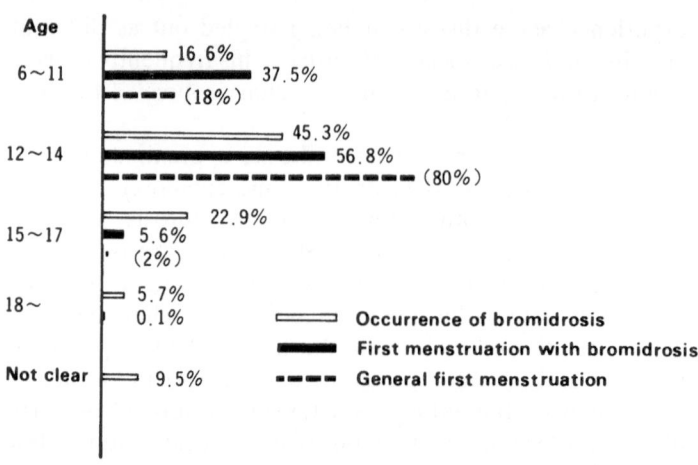

Fig. 8.4. Relationship between the age of occurrence of bromidrosis and that of first menstruation. (Reproduced with permission from Inaba 1986)

8.5 Relationship Between Onset of Menstruation and First Bromidrosis Occurrence in Female Patients

The age at which female patients experience their first menstrual period was collated with the first recognized occurrence of bromidrosis (Fig. 8.4). The better diet and environment available to those children born after World War II has somewhat accelerated physical development, with menstruation occurring at an earlier age. The first occurrence of bromidrosis was likewise somewhat earlier. It is clear that hormonal secretions which lead to menstruation also contribute to the occurrence of bromidrosis. The highest correlation is found among middle (junior high) school pupils.

8.6 Incidence of Bromidrosis Among Male and Female Senior High School Pupils

Table 8.2 shows that among senior high school pupils, the frequency of bromidrosis is higher in females (13%) than in males (8%). The average rate of occurrence is 10%. Both junior and senior high school pupils in Japan become acutely concerned about this condition once they detect it in themselves. Special treatment is sought to put a permanent end to it.

Table 8.2. The frequency of occurrence of bromidrosis among senior high school students (628 males, 442 females)

	Male	%	Female	%	Total	%
Bromidrosis	50	8	57	13	107	10
Nonbromidrosis	578	92	385	87	963	90
Total	628	100	442	100	1070	100

(Reproduced with permission from Inaba 1986)

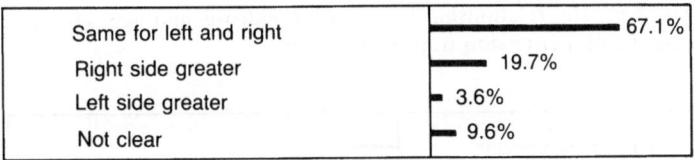

Fig. 8.5. The difference between left and right axillary bromidrosis. (Reproduced with permission from Inaba 1986)

8.7   Differences Between Left and Right Axillary Bromidrosis

Figure 8.5 shows that the levels of bromidrosis in both the left and right axilla are perceived by most patients as about the same (67.1%). However, more of them claimed more offensive odor in the right armpit as opposed to those who claimed this to be true of the left armpit, and these perceived differences may be due to more strenuous use of one arm as opposed to the other. It is worth noting here that almost all Japanese are right-handed; left-handedness is considered unnatural and is strictly discouraged when first detected in most children.

8.8   Relationship Between Bromidrosis and Aging

Figure 8.6 shows a relationship between bromidrosis and the aging process. Over 40% of the patients claimed that the condition had become more offensive, i.e., more conspicuous, with advancing age. Almost 20% claimed the condition was most acute in the beginning, then stabilized to a less acute level. A few patients claimed a decline in the condition with advancing age. Those who were not clear about this accounted for about one-third. Females who had passed menopause generally reported a decline in the condition, indicating a close interaction between bromidrosis and female hormone secretion.

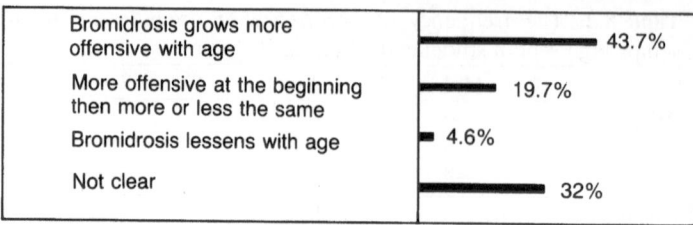

Fig. 8.6. The relationship between bromidrosis and age. (Reproduced with permission from Inaba 1986)

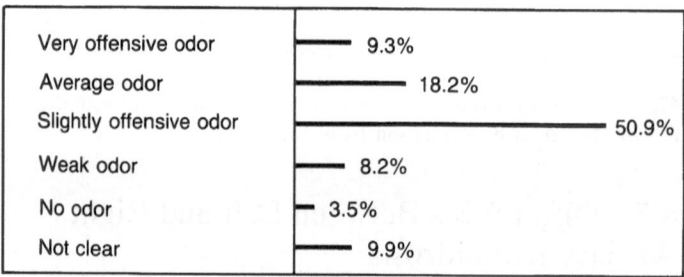

Fig. 8.7. The degree of bromidrosis according to the patient. (Reproduced with permission from Inaba 1986)

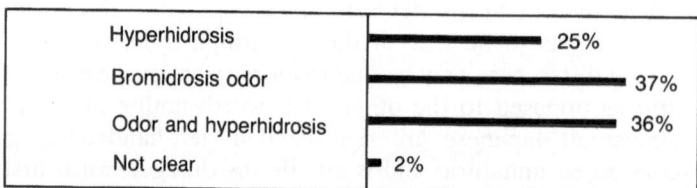

Fig. 8.8. Relationship between anxiety (worry about odor) and which is more serious: hyperhidrosis, bromidrosis odor, or both. (Reproduced with permission from Inaba 1986)

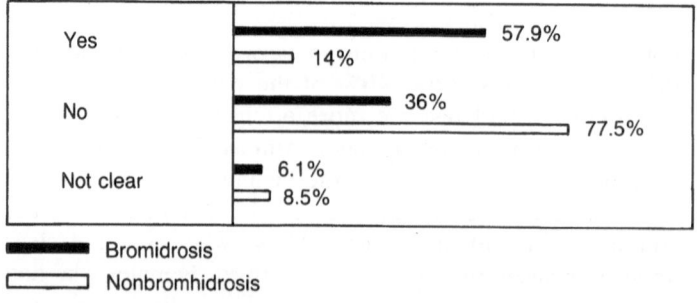

Fig. 8.9. Patients troubled by wearing suits. (Reproduced with permission from Inaba 1986)

8.9 Degree of Bromidrosis by Patient Perception

Figure 8.7 shows patients' evaluations of their own condition by strength of odor, i.e., level of offensiveness to self and others. Essentially the vast majority of bromidrosis patients consider the condition only mildly offensive. However, this information is quite subjective. Many patients do not monitor the odor closely, and it is not possible for the physician, by cursory examination, to determine the strength of the odor, especially when the patient is using a deodorant. This has led many physicians to wonder why patients seek a permanent means of treatment for the condition. The fact is that many of them are not only concerned about the bromidrosis odor but also about the hyperhidrosis condition that discolors clothing. Because the odor can be minimized with regular use of deodorants, many patients are far more concerned about the problem of excessive sweating.

8.10 Anxiety Over Bromidrosis Vs Hyperhidrosis

About one-quarter of bromidrosis patients are concerned more about excessive sweating which discolors clothing than about the odor as such. A higher number show more concern about the odor. More than one-third show more or less equal concern about both odor and hyperhidrosis (Fig. 8.8).

8.11 Patients Concerned About Wearing Heavy Clothing

Generally, patients first think that it is the odor of bromidrosis that distresses them most, but in fact they are concerned both by the bromidrosis odor and the clothing discoloration caused by hyperhidrosis. Physicians may recommend deodorants or electrocoagulation for relief of bromidrosis, but these treatments have no effect on hyperhidrosis. Salesmen who use deodorants may still be embarrassed by shirt stains at the armpits caused by excessive sweating. They try to conceal the condition by wearing a suit jacket. Such heavy overclothing, however, causes body temperature to rise and, in turn, gives rise to more excessive sweating. The shirt is stained more indelibly and must be thrown out (Fig. 8.9).

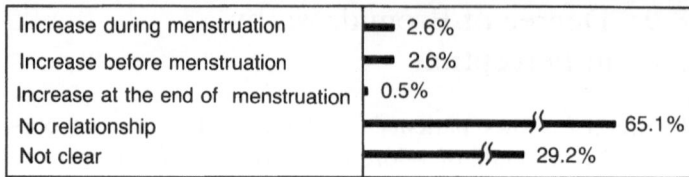

Fig. 8.10. Relationship between bromidrosis and menstruation. (Reproduced with permission from Inaba 1986)

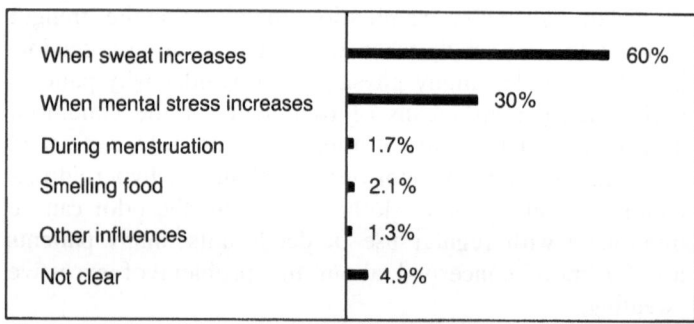

Fig. 8.11. When does the odor develop or increase? (Reproduced with permission from Inaba 1986)

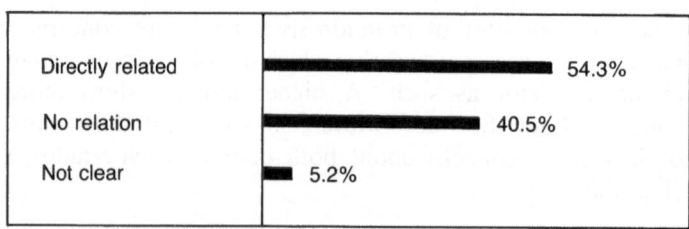

Fig. 8.12. Relationship between bromidrosis and psychosomatic factors. (Reproduced with permission from Inaba 1986)

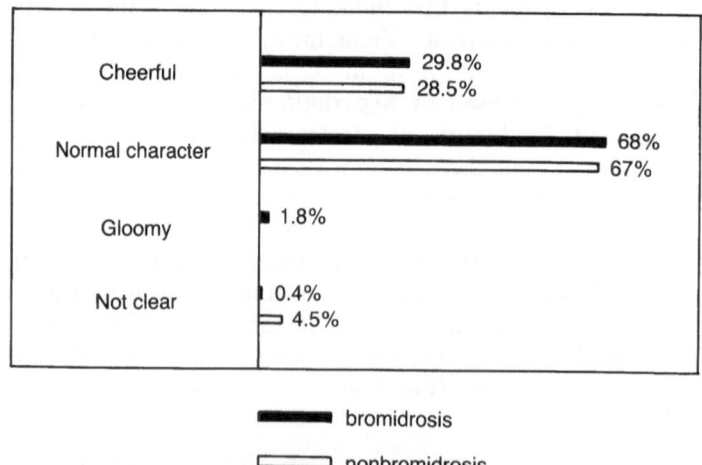

8.12   Relationship Between Bromidrosis and Menstruation

Bromidrosis and menstruation are influenced by hormone secretion, but the relationship between the two is not direct. About 65% of female patients notice no difference in bromidrosis level during menstruation (Fig. 8.10).

8.13   When Does Odor Increase?

In cases of bromidrosis, there is a clear correlation between increased bromidrosis and increased sweating or increased mental stress (Fig. 8.11).

8.14   Relationship Between Bromidrosis and Psychosomatic Factors

As shown in Fig. 8.12, sweat excretion can be influenced by psychosomatic factors (54.3%). During a school examination, for instance, secretions are usually increased. Concern over bromidrosis odor can trigger psychosomatic stress that results in a higher degree of hyperhidrosis, thus establishing a vicious cycle.

8.15   Mental State of Bromidrosis Patients

A comparison of the self-assessed mental states of bromidrosis and nonbromidrosis patients (Fig. 8.13) shows that those with a cheerful disposition account for less than 30%, respectively. Those in the average range account for more than two-thirds, respectively. Bromidrosis patients with a "dark" disposition are few in number, as are those 0.4% in the "not clear" category. These findings indicate no proportional differences in general mental state between the bromidrosis and nonbromidrosis groupings.

Fig. 8.13. Mental state of bromidrosis patient: character/behavior. (Reproduced with permission from Inaba 1986)

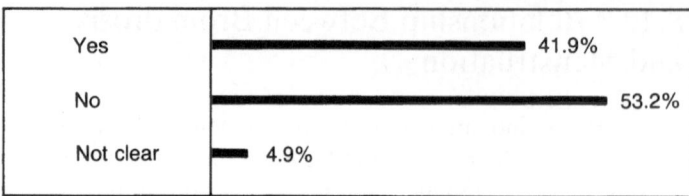

Fig. 8.14. Does bromidrosis influence the character of the patient? (Reproduced with permission from Inaba 1986)

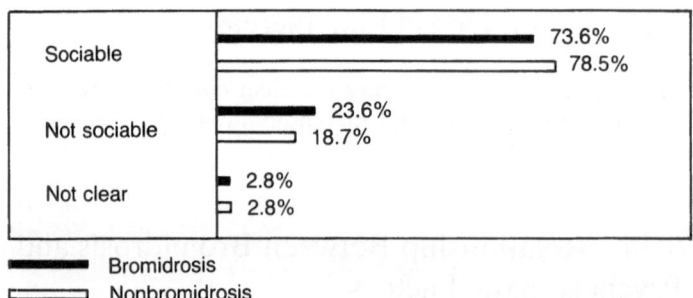

Fig. 8.15. Sociability of the bromidrosis patient. (Reproduced with permission from Inaba 1986)

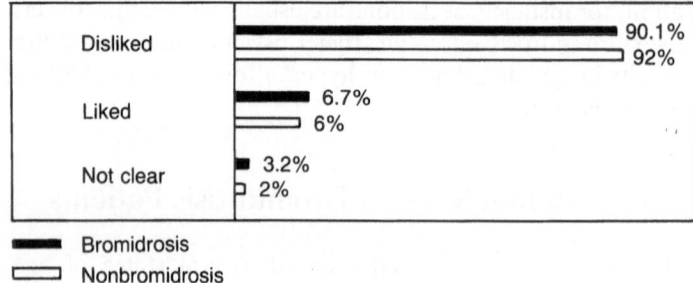

Fig. 8.16. Dislike of bromidrosis odor. (Reproduced with permission from Inaba 1986)

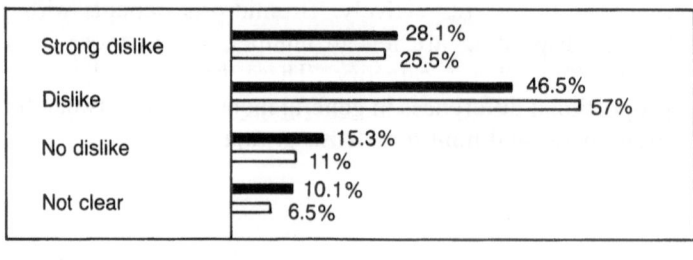

Fig. 8.17. Attitude of both bromidrosis and nonbromidrosis groups toward cases of bromidrosis. (Reproduced with permission from Inaba 1986)

8.16   Does Bromidrosis Influence Behavior?

Figure 8.14 shows the relationship of bromidrosis to the patient's behavior. Bromidrosis patients who say their personal behavior is influenced by the condition account for 41.9%, those who claim no influence 53.2%, and those "not clear" 4.9%. In some cases, young patients do not wish to attend school because the condition will be detected by other students and the patients will be subjected to ridicule.

8.17   Sociability of Bromidrosis Patients

Sociability or nonsociability of bromidrosis patients (73.6%) and nonbromidrosis patients (78.5%) seems to be almost equal. However, it is also true that bromidrosis patients take extra pains to conceal the condition in social situations. When going on a trip with friends, for instance, these patients may hesitate to take a communal bath with others for fear the odor will be noticed, or may wish to apply a deodorant to the armpits surreptitiously. This concern may reach a degree that discourages these people from involvement in certain social activities, and thus lessens close friendship when a "risk factor" is involved (Fig. 8.15).

8.18   Dislike of Bromidrosis Odor

The odor of bromidrosis is disliked by more than 90% of both the bromidrosis and nonbromidrosis groups, although 6% report a liking for it, and only 2%–3% in these groups are "not clear" about it. The fact that most people dislike the odor is probably true of all advanced cultures in which deodorants are used to conceal it (Fig. 8.16). In Japan, the odor is offensive to most people regardless of their economic class and social standing.

8.19   Attitudes of Bromidrosis and Nonbromidrosis Groups Toward Bromidrosis

Both the bromidrosis and nonbromidrosis groups mostly dislike other people who have bromidrosis, with the degree of dislike considered intense in 25% or so of both groups. However, almost one-quarter of those in both groups claimed no dislike or "not clear" attitudes toward people who have bromidrosis (Fig. 8.17).

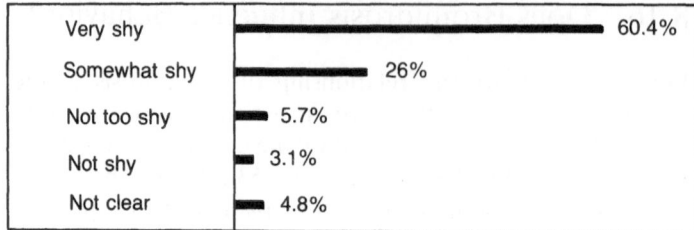

Fig. 8.18. Shyness about bromidrosis odor. (Reproduced with permission from Inaba 1986)

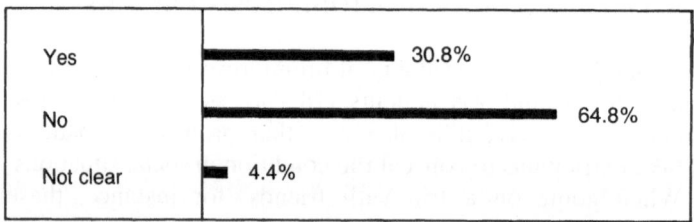

Fig. 8.19. Percent of bromidrosis patients who learn about the condition from other persons. (Reproduced with permission from Inaba 1986)

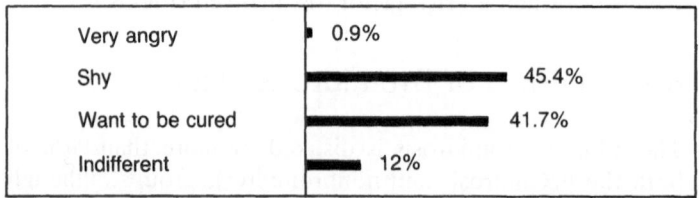

Fig. 8.20. Reaction of bromidrosis patients to their condition. (Reproduced with permission from Inaba 1986)

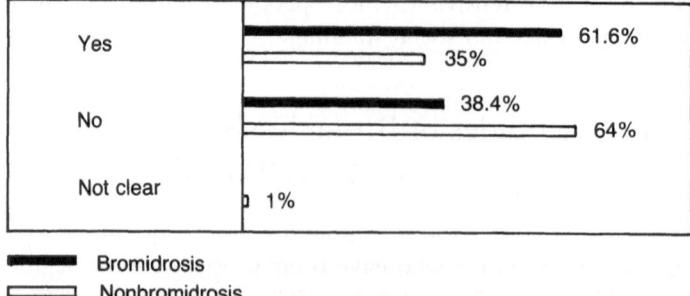

Fig. 8.21. Whether hyperhidrosis occurs on palms and soles during mental strain. (Reproduced with permission from Inaba 1986)

8.20 Shyness About Bromidrosis

A significant 60.4% of bromidrosis patients feel very shy about the condition, about one-quarter feel somewhat shy, and the remainder do not feel shy or are "not clear" on this point. The finding that 86.4% do feel very or somewhat shy about the condition indicates their high degree of concern about it in a society in which the condition is relatively uncommon (Fig. 8.18).

Among married couples, husbands whose wives have the condition will not comment directly on the odor but will demand that the wife take a bath before lovemaking. This causes the wife to lose sexual desire due to anxiety about the odor.

8.21 Patients Who First Learned About Their Condition from Other People

Almost one-third of bromidrosis patients first learned of their condition from other people (family, friends, etc.), whereas those who were first aware of it themselves account for the majority. Fewer than 5% were "not clear" on this issue. The fact that almost one-third were informed of their condition by others is of interest; in Japanese society people are usually very reticent about commenting on such matters directly to the person concerned (Fig. 8.19).

8.22 Reaction of Bromidrosis Patients to Their Condition

Reactions of bromidrosis patients toward the condition can differ somewhat. A very small number said it made them feel very angry; a much larger number said it made them feel shy. The chief reaction of about 40% was a concern to find a cure for the condition. Only a small number claimed to be indifferent to it. Most of these patients thus show some form of acute concern about the condition, some to the point of neurotic fixation on it (Fig. 8.20).

8.23 Mental Stress and Occurrence of Hyperhidrosis on Palms and Soles

There is a significant correlation between mental stress and occurrence of excessive sweating on the palms of the hands

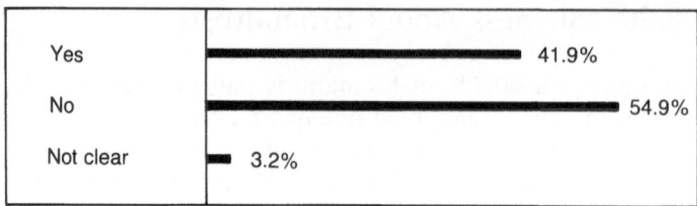

Fig. 8.22. Whether a close relationship was ever lost because of the condition. (Reproduced with permission from Inaba 1986)

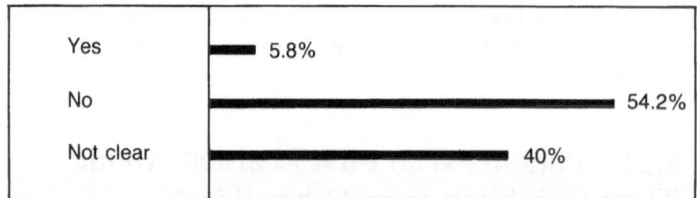

Fig. 8.23. Was there a clear relationship between such loss and the bromidrosis? (Reproduced with permission from Inaba 1986)

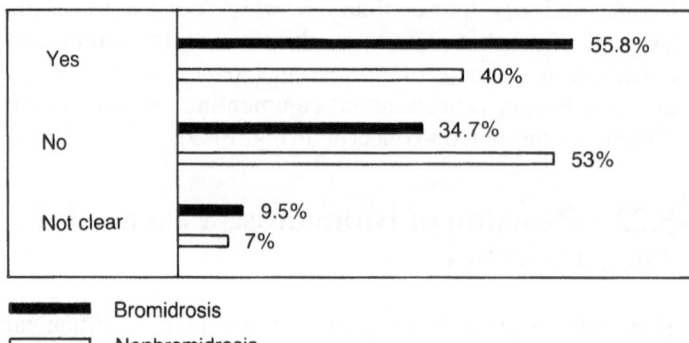

Fig. 8.24. Relationship between bromidrosis and disturbed marriage cases: 300 bromidrosis patients and 200 nonbromidrosis. (Reproduced with permission from Inaba 1986)

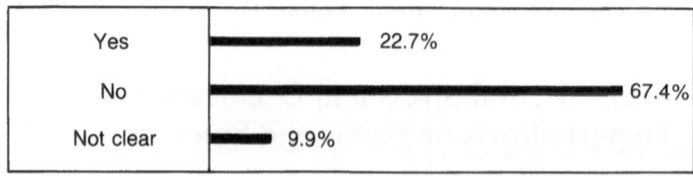

Fig. 8.25. Relationship between bromidrosis and thoughts of committing suicide. (Reproduced with permission from Inaba 1986)

or the soles of the feet. The point here is that hyperhidrosis in both groups is not necessarily correlated with bromidrosis, although far more bromidrosis than nonbromidrosis cases are prone to hyperhidrosis (Fig. 8.21).

8.24 Presumed Loss of a Close Relationship Due to Bromidrosis

Although a significant number presumed the loss of a close friendship—particularly with a member of the opposite sex—due to the condition, more than half had not had that experience. Only a few were "not clear" one way or the other (Fig. 8.22).

Many patients do presume, rather than actually know, that the lost relationship could be traced to their bromidrosis. Only a small number could claim to be definite on this point, whereas more than half felt it was not perfectly clear. Another 40% were "not clear" one way or the other (Fig. 8.23).

8.25 Bromidrosis as a Cause of Marital Problems

More than half of married bromidrosis patients claimed marital problems which were directly attributable to the condition. Another 40% of nonbromidrosis cases claimed this to be true of marriage to a bromidrosis partner. On the other hand, about one-third of the bromidrosis group and over half of the nonbromidrosis group claimed the condition had not caused any marital problems. Less than 10% in both groups were "not clear" on this issue (Fig. 8.24).

Married women feel the most concerned to conceal the condition so that it does not give rise to partner dissatisfaction. Even when the condition is noticed, however, many partners, especially males, become habituated and therefore indifferent to the odor. Nevertheless, bromidrosis does account for trouble in a comparatively large number of Japanese marriages in which it is a factor.

8.26 Bromidrosis and Suicidal Thinking

This finding indicates that less than one-quarter of bromidrosis cases are concerned about the condition to the point of suicidal thinking. The vast majority have learned "to live with it," albeit at times with great anxiety. Less than one-tenth of the bromidrosis cases were "not clear" on this point (Fig. 8.25).

It is still significant, however, that almost one-fourth of these people are seriously distressed by the condition to the point of suicidal depression, which indicates the gravity placed on the problem in Japan. It does not seem to generate so acute a concern in the Western world, where it is far more common and therefore seemingly less mentally stressful. When this is not the case, obsession tends to set in.

Chapter 9.   Body Odors and Associated Diseases and Disorders

9.1   Genital Bromidrosis

Human vaginal odor most likely consists of a number of components, including ordinary odor in the genital region (vulvar secretions from sebaceous, sweat, Bartholin's and skene's glands), mucus from the cervix, endometrial and oviductal fluids exuded from the vaginal walls, and exfoliated cells of the vaginal mucosa. Some female bromidrosis patients refer to malodor in the genital region, but do not know whether it is genital skin surface bromidrosis or due to vaginal discharge. Some report discolored underclothing. These patients are concerned about the malodor as well as the discoloration, and wish to receive permanent treatment relief.

Apocrine glands are present in the pubic hair region appended to pubic hair follicles. This indicates there can be skin surface bromidrosis in this region. The condition can be detected from the pubic hair region during uterine curettage. In one study of 855 patients, 80 (9.4%) were found to have axillary bromidrosis. Among them 23 (29%) also had pubic zone bromidrosis, indicating that 2.7% out of the total study group had bromidrosis in both the axillary and pubic areas (Inaba et al. 1973). Another study of 232 axillary bromidrosis cases examined by a questionnaire revealed that 26 (11.2%) had pubic bromidrosis odor as well. Since the frequency of bromidrosis averages 10% in the Japanese population, pubic bromidrosis can be seen to average 1.12%. It is estimated that among the Japanese female population as a whole, an approximate 1.12%–2.7% have pubic bromidrosis.

Mitsumine and Okugawa (1940) reported on 92 cases of axillary bromidrosis, ranging in age from 20 to 50 years, with 31 (33%) revealing apocrine glands in the pubic area

under histological examination. Yoshihiro (1942) reported
that in 2 cases among 112, the pubic bromidrosis odor was
quite similar to the odor of axillary bromidrosis. In Western
nations where the diet is high in calories and animal fats,
the percentage of pubic bromidrosis is predictably higher
than in Japan.

In investigating the relationship between pubic bromidrosis and menstruation, questionnaire data showed that
pubic bromidrosis was apparently related to menstruation in
27% of the cases but was unassociated in 73%. Within that
27%, the pubic bromidrosis odor during menstruation is
even stronger than usual (Inaba 1986). This may be due
to the combination of pubic bromidrosis odor with that
of vaginal discharge. Genital bromidrosis odor increases
during pregnancy and is particularly strong during the late
stages of pregnancy. This is a result of hyperfunction of the
apocrine glands due to hormonal influence.

We have found no direct relationship between the strength
of axillary bromidrosis odor compared to that of pubic bromidrosis. Differences observed are individual differences.
Kligman and Shehadeh (1964) reported a study of apocrine
glands in the pubic region to determine why this area lacks
the odor characteristic of the axillae. The glands were found
to be anatomically perfect but physiologically functionless.
As such, they cannot produce apocrine sweat. The pubic
apocrine gland is said to be further evidence for the physiological decline of these structures among the higher primates. In other words, this gland is en route to extinction.

Other genital odors unrelated to genital bromidrosis may
include normal discharge from the vagina, the odor of
menstrual blood, and pathological odors such as those associated with cancer. Adult women normally secrete fluor
albus which usually smells less unpleasant. Vaginal odors
are slightly less pleasant and more intense during the
menstrual cycle. Although average secretion odor is somewhat less pleasant in the preovulatory and ovulatory phases,
considerable variation is observed in individual cycles (Doty
and Huggins 1975). Any or all of these components may
contribute substrates for bacterial production of volatiles.

Lactic acid and 3-hydroxy-2-butanone, as well as C2 to
C5 aliphatic acids, have been identified, suggesting the presence of volatiles produced by microbial action (Michael
et al. 1974; Preti and Huggins 1975). The odor of vaginal discharge depends on the quantity of Doderlein's bacilli. Chemical identification of vaginal secretions has revealed a variety
of low molecular weight organic compounds. Lactic acid and
short-chain aliphatic acid may be present (Labows 1979).

Young women who do not cleanse their genitals regularly may find sebum (smegma) around the genital area. After a few days the smegma increases in quantity and, in combination with bacterium, produces odor.

The odor of menstrual blood as such is typical. The discharge of women who have bacterial vaginosis often has a fishy odor. This odor may be intensified by adding a strong base to the vaginal discharge, indicating trimethylamine, the substance that causes the characteristic putrid odor in fish (Brand and Galask 1986).

One report (Brand and Galask 1986) states that all patients with a detectable fishy odor had a vaginal pH of 5–6 and readily identifiable amounts of trimethylamine. All of the normal patients had a vaginal pH of 4.5 or less and no detectable trimethylamine in gas chromatographic analysis. Although the fishy odor and the presence of trimethylamine in vaginal discharge from patients with bacterial vaginosis have been confirmed, the biosynthetic origin of this substance has not been clarified. Trimethylamine produced by reduction of trimethylamine oxide in fish muscle (Barett and Kwan 1985) is responsible for the putrid odor.

Certain diseases are secreted by malodorous vaginal discharge. Vaginal infection connected with the Gordmerella vaginalis organism is commonly called "non-specific vaginitis" characterized by a malodorous vaginal discharge (Pheifer and Forsyth 1978; Meech and Loutit 1985). Anaerobic bacteria involved in producing volatile fatty acids and amines cause the malodor (Speigel et al. 1980; Chen et al. 1979). In cases of cancer, Doderlein's bacilli quantity is decreased, and vaginal discharge contaminated by other bacterium reveals a stronger odor. If a sanitary napkin or tampon is not replaced for a few days, the vaginal odor is increased due to bacterial contamination of vaginal discharge.

9.1.1 Fluor Test Method

Inaba et al. (1979c) invented a device for simple home examination of the condition of vaginal discharge (Fig. 9.1.a,b). This fluor albus test (the Fluor test) relies on a cytocellophane method, and is capable of auto-collecting vaginal content and determining pH values and the state of cleanliness. The technique has been applied to the smear test and its feasibility as an auto-collecting technique has been studied. Since cytodiagnostic studies are currently performed over a wider scope of applications, including inflammatory, endocrine, and oncological abnormalities,

a

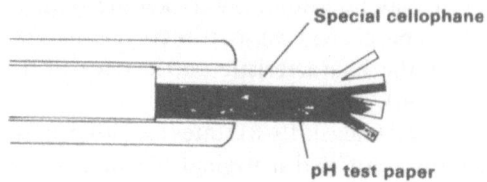

b

Fig. 9.1.a, b. The Fluor test method (cytocellophane method) is capable of auto-collecting vaginal content and determining pH values and state of cleanliness. **a** Fluor tester (auto collection unit). (Reproduced with permission from Inaba et al. 1979c) **b** Structure of the Fluor tester. It is necessary to use a sampling tool which will not allow pH values to fluctuate due to Bartholin secretions. A spring is attached around the sampling rod inside the Fluor tester. MR-BCG test paper and special cellophane are coiled around the distal tip of the sampling stick. The distal tip of the stick is split like a V. Following insertion of the tester into the vagina, the piston is moved to thrust itself out of the sampling stick. Sampling of vaginal discharge is done while this stick is rotated. (Reproduced with permission from Inaba et al. 1979c)

the usefulness of the Fluor test method merits recognition in this respect, as well as for its traditional role of determining the state of vaginal discharge.

Whether the leukorrheal status is favorable or not depends on the volume of lactobacilli (Doderlein's bacilli). These bacilli function to keep the vagina clean. According to the established standard, the vaginal pH value of a healthy, normal female is in the range of 4.0–4.8, and with vaginal or associated disease, 5.6 or greater. When the pH value is between 5.0 and 5.4, further examination of vaginal content is undertaken with consideration of macroscopic findings and the state of vaginal cleanliness, as well as menstruation and the puerperium stage.

To ensure that the Fluor tester can differentiate pH values easily, a pH test paper called MR-BCG (Toyo Roshi Co., Ltd., Tokyo) was developed by coupling MR (methyl red; pH 5.4–7.0) and BCG (brom cresol green; pH 4.0–

Table 9.1. Determination of the pH value is done immediately after wiping the clear section of the Fluor tester by means of color showing on the tip of a particular pH test paper inside the tester. It is also done by pushing the piston, which protrudes the tip of the special pH test paper, out of the Fluor tester. (Reproduced with permission from Inaba et al. 1979c)

Bacterial content	Degree I	Degree II	Degree III
Color of the particular test paper	Pink	Purple	Blue
pH value	4.0~4.8	5.0~5.4	5.6 or above
Status of fluor (leukorrhea)	Normal	Somewhat normal	Treatment is necessary

5.6) so it is capable of indicating pH by color in the range of pH 4.0–7.0. Secretions permeate and climb by capillary action along the V-shaped distal tip of the special cellophane wound around the sampling stick. The sampled material is fixed with 95% alcohol. The state of cleanliness is then determined using Gram's stain, which indicates the presence or absence of lactobacilli or other miscellaneous bacteria. The state of cleanliness is classified as follows:

1. Degree I—only lactobacilli are present
2. Degree II—lactobacilli and other miscellaneous bacteria are found to be present in a mixed condition
   (a) More lactobacilli than other miscellaneous bacteria
   (b) Intermediate between (a) and (c)
   (c) More miscellaneous bacteria than lactobacilli
3. Degree III—only other miscellaneous bacteria are present.

The following general conclusions were reached by using the Fluor tester clinically. First, the tester is capable of determining the status of vaginal content easily by means of the MR-BCG test paper inside the tester. Second, the same results were obtained with regard to sterility of vaginal content, and Papanicolaou's stains from samples attached to the cellophane coiled around the sampling tool were obtained from smeared samples. Third, favorable results were achieved by applying the Fluor tester not only as an auto-sampling method, but also for the collection of smears by rubbing the uterine cervix. Fourth, in Papanicolaou's stain, the smear test can be useful in detecting inflammation or endocrinic disorders.

The clearness of the vaginal discharge depends on the amount of Doderlein's bacilli present. This bacillus is the same as that used in making yogurt. The normal pH of vaginal discharge is acid pH (4.0–4.8) (Table 9.1). If the amount of Doderlein's bacilli decreases, the pH of the vagina will increase. If the pH of the vaginal discharge is 5, then the color of the cellophane changes to purple. If it is lower than 5 the color turns pink and on the base side turns blue. By using this simple device, the condition of vaginal secretion can be ascertained and the presence of a disorder or disease readily detected.

9.2   Male Odor

According to Takagi (1988), males have their own genital-associated odors. For instance, L-pyrroline, a steroid having a musk scent (copentadecalactone), is present in semen and 5α-Androst-16-en-3-one is present in urine. Male musk deer attract females by the odor emitted from their sex glands. Human beings became aware of its attractive fragrance and began to use it in cosmetics, which then drove the species to the verge of extinction. The pheromone odor observed among many species of animals is activated under an extremely subtle motivation toward which they seem to show reflex actions by instinct. It is probably the same with the male and female odors that, even though they are very weak, work exactly as they should.

During World War II, an incident was recorded in a female prison in Indo-China. The prisoners made noises once in a while but the reason was not clear at first. Through investigation, it was revealed that they became clamorous when men walked outside the high prison walls. This incident shows that the "male" odors definitely stirred up the female prisoners. A similar phenomenon was observed in fishing boats in the Arctic Ocean where the men engaged in fishing were obliged to live together in a confined area for a long period of time. They soon grew restless and irritable, and often came to blows. In a peaceful civilian society, men and women live together, constantly smelling the "odors" of the opposite sex, whether at home or on a crowded train, and thus their mental state is kept well-balanced.

The vaginal secretion of a female monkey during the mating season has been tested to show that it contains several kinds of low fatty acids. This secretion, or the low fatty acids combined in the same proportion, will induce male sexual behavior when it is smelled by a male monkey of the same species as well as by a chimpanzee, which is a different species.

In a condition of complete phimosis, the glans remains unexposed and smegma collects between the prepuce and the glans. This secretion of sebaceous glands (preputial gland), specifically the thick, cheesy, ill-smelling secretion consisting principally of desquamated epithelial cells, is found under the male prepuce, and comes from Tyson's glands (Edward Tyson, English physician and anatomist, 1649–1708). The modified sebaceous glands located on the penis at the neck and inner surface of the prepuce are similar to those found under the labia minora around the glans clitoridis. Smegma bacillus is termed *Mycobacterium smegmatis*, and is primarily a fast-growing acid-fast bacterium of an antipathogenic characteristic, but it begins to emit a peculiar and intense odor when it is contaminated by ordinary bacteria under unsanitary conditions. Treatment for false phimosis involves exposure of the glans for the purpose of washing and keeping it clean. For genuine phimosis, phimosiectomy is generally performed to avoid smegma accumulation between the glans and prepuce.

9.3 Bromidrosis and Skin Diseases

9.3.1 *Fungal Disease* (Trichosis Axillaris)

A fungal disease sometimes detected in the axillary region causes hyperhidrosis and discolors the hair shaft. The fungi responsible include *Nocardia tenuis*. Yoshihiro (1942) reported 35 cases (18.32%) of this kind among 191 patients treated for bromidrosis and an additional 14.18% among nonbromidrosis patients, all with fungal infections caused by *Nocardia tenuis*. This indicates that bromidrosis patients are rather more susceptible to this type of infection.

Nodules composed of bacterial masses develop in the axillae and sometimes in the pubic hairs in cases of trichomycosis. This condition, which is common in many populations, is caused by diphtheroid *Corynebacterium tenuis*. The infection is abetted by axillary sweating and insufficient hygiene. Discolored stains of yellow, red, or black eventually reveal the infection. The infected hairs fluoresce under Wood's light, and the nodular granules can be observed with a hand lens. In one report (Crissery et al. 1952), this infection was detected in 23 of 100 consecutive patients examined in the United States.

For microscopic examination, the infected hairs are placed on a slide in a drop of 10% KOH under a coverslip. After gentle heating for cleaning, the preparation is examined under a light microscope, revealing that the nodules on the hair are composed of short bacillary forms. The hairs of the infected area are shaved and antibacterial soaps and creams are applied in the treatment procedure.

9.3.2   Hidradenitis suppurativa

An inflammatory process called *Hidradenitis suppurativa*, which is usually recurrent, most often develops in areas that have apocrine glands such as the axillae, the breasts (especially the nipple areolae), and the genital region. Obstruction of the hair canal and staphylococcus inflammations give rise to a heat rash which characterizes this disease. It is frequent in Caucasians but rare in Japanese. A fistula similar to a hemorrhoid develops and can be very resistant to conventional treatment.

9.3.3   Fox-Fordyce Disease

Fox-Fordyce disease is characterized by chronic, pruritic, and widespread eruptions in apocrine areas, principally the axillae and pubes, as a result of obstruction and rupture of the intraepidermal portion of the ducts of the affected apocrine glands. Secondary changes occur in the secretory tubule and adjacent dermis. This disease was first described in 1902 when Fox and Fordyce reported two cases to the New York Dermatological Society (Goodman 1926). Unique to females, it begins to develop in the apocrine glands during puberty. It is engendered by a heat rash caused by retention of sweat in the apocrine glands.

Four types of tumors are associated with the apocrine sweat glands: apocrine nevus, apocrine duct cyst, apocrine gland tumor, and apocrine gland cancer. However, none of them are related to a condition of bromidrosis.

9.3.4   Eczema

A condition of eczema is caused by excessive sweat gland excretion and an increase in pH level. It is most common among obese patients, subject to contact dermatitis in the axillary region. Another causal factor is skin irritation due to excessive use of deodorants in cases of bromidrosis. Severe cases, such as nevus, may result in skin pigmentation turning reddish, brownish, or black.

9.4   Halitosis

Halitosis is not usually the result of a severe medical problem, but it is often a social drawback. Few people wish to associate closely with anyone who emits a foul mouth odor. People who work closely with the public must be especially concerned in dealing with this condition. In numerous cases the individual is unaware of the condition until informed of it by others. There is, however, another perplexing group of patients who complain of having halitosis but present no objective evidence to substantiate their claims.

Quality and intensity of breath normally changes with advancing age. The breath of infants and children ordinarily seems to be sweet and pleasing. In adolescents the breath is rather heavy and a bit pungent but not unpleasant. In middle age, however, the breath is characteristically less pleasant, even with fastidious oral care. With more advanced age the breath is usually heavy, pungent, rather sour, and so intense that it may be unpleasant even if oral hygiene is meticulous. Why this occurs is unclear, but it may be associated with both local and systemic factors (Sulser et al. 1939).

In describing the etiology of halitosis, Claycomb and Schearer (1986) noted that approximately 90% of mouth odors come from the oral cavity. Healthy mouths can emit foul odors due to odoriferous foods such as garlic, consumption of alcohol, tobacco smoking, an imbalance in oral microorganisms, and/or a decrease in salivary flow. Salivary flow is decreased during sleep. Odor in the morning is more intense for those who breathe through the mouth or snore while asleep, but it will not persist after oral hygiene or even after drinking and eating.

Normal saliva has an ordinary pH of approximately 6.5, a slightly acidic pH which suppresses the growth and proliferation of gram-negative and anaerobic bacteria. On the other hand, saliva removed from the mouth and incubated rapidly turns alkaline, favors gram-negative bacteria, and allows activation of enzymes necessary to the putrefaction of amino acids (the end products of which are the malodorous sulfur-containing compounds). Overnight putrefaction of saliva in the mouth (as a warm, moist oral incubator) can produce odors of very offensive strength even if the individual emits no foul odor during the daytime (Massler et al. 1951).

Local conditions that may lead to halitosis include poor oral hygiene, numerous caries, gingivitis, periodontitis, and other conditions that permit food impaction, Vincent's infection, hairy or coated tongue, fissured tongue, excessive smoking, healing extraction wounds, and necrotic tissues due to ulceration. In adults, chronic periodontal disease is a major cause, with periodontal pockets producing hydrogen sulfides which have an offensive odor and are also involved in trapping food. Periodontal therapy may alleviate the condition of halitosis by diminishing volatile sulfur compounds.

Halitosis may also be associated with an increase of gram-negative filamentous organisms or an increase in pH to 7.2 and may be associated with indoles and amines formed in the oral cavity. Another cause is a dentigerous cyst with a fistula draining into the oral cavity. Other causal conditions may include abcesses, nasal tumors, chronic sinusitis with a postnasal drip, rhinitis, lethal granulomas, pharyngitis, tonsillitis, syphilitic ulcers, cancrum oris, ulcerogangrenous processes, cancerous tumors of the trachea and bronchi, chronic fetid bronchitis, and infectious malignant neoplasms of the oral and pharyngeal cavities (Lu 1982).

9.4.1   Oral Diseases and Malodor

Pyorrhea. Among the oral diseases that cause malodor of the mouth, pyorrhea ranks the highest in the rate of incidence. It is a disease of the gum and the texture surrounding teeth and can be defined as the advanced stage of gingivitis when the inflammation extends as far as the cusp or dental alveoli. The common symptoms of pyorrhea in its early stage are the swelling of and bleeding from the gum. In the more advanced stage, the teeth become unstable and then finally fall out. The foul odor is caused by the generation of pus due to the inflammation. Treatments vary according to each stage of the disease but the principal care would be instruction on keeping the oral cavity clean and the removal of odontolith by specialists.

Gingivitis. This inflammation of the gum develops as a result of the formation of sordes or odontolith. Because the human oral cavity has an ideal temperature and moisture for bacterial propagation, malicious bacteria grow to cause gingivitis under foul conditions which will finally lead to pyorrhea. Mouth odor resulting from gingivitis precedes the one from caries in the frequency of occurrence. For prevention, regular removal of odontolith by dentists is essential.

Caries. Caries are formed by a bacterium called *Streptococcus mutant. Streptococcus mutant* converts the sugar contained in the residual food adhered to tooth surfaces into acid which then forms caries in the tooth. When caries develops as far as the nerve pulp, the pulp decays and forms pus. This generation of pus causes a foul odor called "gangrenous odor" which is produced as a result of protein decomposition. Regular brushing of teeth is naturally recommended for the prevention of tooth decay.

Hairy tongue. Magnification of the tongue surface by a microscope reveals a very fine uneven feature, each minute protrusion of which is called a papilla. Residual food and the old cells continuously being shed from the oral cavity surface easily accumulate over this uneven tongue surface and cause a malodor of the mouth. The greater the quantity of the cell components in saliva the more acute the mouth odor will be due to the decomposition of such substances as amino acid contained in saliva producing a putrid smell. Various bacteria may propagate over this unsanitary area and the tongue surface appears as if it were covered by a very thin layer of moss with the color varying from white to black. This is called a "hairy" or "coated" tongue. As the accumulation gets heavier, the overgrowth of bacteria causes a more putrid smell. For sanitary purposes, the tongue surface requires regular brushing by a specially designed brush.

9.4.2 Halitosis Due to Systemic Factors of Pathologic Origin

The other 10% of mouth odors can be ascribed to lung-borne systemic sources. Lung-borne malodors are detectable by instructing the patient to breathe out through the nose while keeping the mouth closed. If a malodor can still be detected, the patient should accordingly be referred to a specialist, since a pathological metabolic disorder may be indicated. These odors are more intense and permanent, typically including ammonia accompanying uremia, acetone for the poorly controlled diabetic, and pulmonary disease (Tonzetich 1978).

Lu (1982) points out that diabetes is the best-known example of halitosis from a systemic condition of pathologic origin. Although the odor cannot be detected in well-controlled patients, an acetone-like, sweet, fruity odor indicative of diabetes acidosis or impending hyperglycemic coma may occur, due to abnormal blood accumulation of ketones excreted through the respiratory system. Ammonia and urine odors detected in the breath may indicate uremia or kidney failure. In cases of severe hepatic failure the character of the breath, called "fetor hepaticus", produces a sweetish, feculent, "amine" odor similar to that of a fresh cadaver. This kind of breath often precedes a condition of hepatic coma. It may also be present in a patient with extensive portosystemic anastomosis, but by its nature is intermittent for a long period of time. An acid sweet odor can indicate acute rheumatic fever. A foul breath odor similar to putrefying meat can suggest a lung abscess or bronchiectasis due to the dilatation of pus-secreting bronchi.

Other systemic diseases that give rise to halitosis include lung gangrene and pulmonary tuberculosis. It can occur in cases of toxemia, gastrointestinal disorders, or hemorrhage at any portion of the gastrointestinal tract. It can also exist in neuropsychiatric disorders in which patients complain subjectively only of "bad breath." These particular diseases produce halitosis to some extent but the condition is aggravated by poor oral hygiene common to patients who neglect oral hygiene due to physical or mental stress.

Emotional upsets that affect digestion and body chemistry may occasionally have an influence on the breath. Patients who suffer from nonlipid reticuloendotheliosis disorders, such as eosinophilic granuloma, Letterer-Siwe disease, and Hand-Schuller-Christian disease, are prone to complain of halitosis, sore mouth, and a repugnant taste.

Cases of acute and chronic scurvy due to vitamin C deficiency almost always reveal the typical foul breath associated with fusospirochetal stomatitis. In cases of macroglobulinemia, primary herpes simplex infection, hemophilia, von Willebrand's disease, cryoglobulinemia, aplastic anemia, polycythemia vera, agranulocytosis, leukemia, infectious mononucleosis, thrombocytopenic purpura, and thrombocythemia, halitosis commonly occurs due to infection, necrosis, and blood decomposed after spontaneous bleeding in the oral cavity.

Patients afflicted with noma show the typical, if more intense, breath with odor of acute necrotizing gingivostomatitis. A rapidly disseminated gangrene of the oral and facial tissues, noma occurs mostly in patients who are debilitated or undernourished and in cases of diphtheria, dysentery, measles, pneumonia, scarlet fever, syphilis, tuberculosis, or blood dyscrasias. Even though noma primarily occurs in the oral maxillofacial area, it can be considered a secondary complication of a systemic disease.

9.4.3   Biochemical Research on Halitosis

Halitosis does have a multifactorial etiology, but local factors assume the major role in most of the cases caused by bacteria and substances which contain or are capable of producing hydrogen sulfide, dimethyl sulfide, methylmercaptan, and ethyl mercaptan. Procedures that serve to reduce these bacteria and substances will suffice to eliminate most of the causes of halitosis.

Richter and Tonzetich (1964) detected such volatile sulfur compounds (VSC) as hydrogen sulfide (H_2S), methylmercaptan (methyl sulfhydrate [CH_3SH]), ethyl mercaptan (C_2H_5SH) and dimethyl sulfide (($CH_3)_2S$) from putrid saliva emitting bad odors. Microorganisms in salivary sediment degrade as the proteinaceous component of sloughed cells and primarily produce H_2S + CH_3SH, a fact which links VSC with halitosis. In exploring the genetic mechanism of VSC, Tonzetich and Richter (1964) incubated whole saliva, compared it with saliva supernatant, and concluded that VSC production occurs due to rupture of cellular constituents in whole saliva. Tonzetich (1978) also detected a VSC increase and intense odor in the mouth air of individuals suffering from inflammatory conditions. Concentrations of VSC are known to increase with advancing age.

Rizzo (1967) reported possible detrimental effects of H_2S on gingival tissues after discovering H_2S present in all pockets larger than 2 mm in size. Nag (1984) conducted an in vitro study of porcine sublingual mucosa, and found that the CH_3SH and H_2S volatile thiols increased mucosa permeability, a condition which could be reversed by application of thiol reacting reagents.

It is very difficult to judge how one's own breath smells. The sense of smell is the most unreliable of the senses and has still not been scientifically clarified. Various instruments have been developed to detect halitosis, and the relation between halitosis and offensive substances in the oral cavity has been studied. Gas chromatographic analysis was applied in the 1960s. Weber and Rhoades (1960) detected acetone, methanol, and ethanol in saliva; Eriksen and Kulkarni (1960) detected methanol in saliva; and Tonzetich (1971) detected hydrogen sulfide, methylmercaptan, and methyl sulfide in the breath. The 1970s witnessed a further development of analyzing detectors, and the flame photometric detector (FPD) was added to the conventional thermal conductivity detector (TCD) and flame ionization detector (FID) in gas chromatographic analysis. This enabled researchers to analyze phosphor and sulfuric compounds

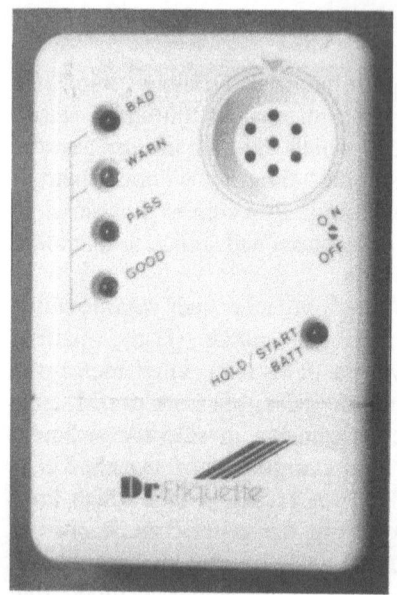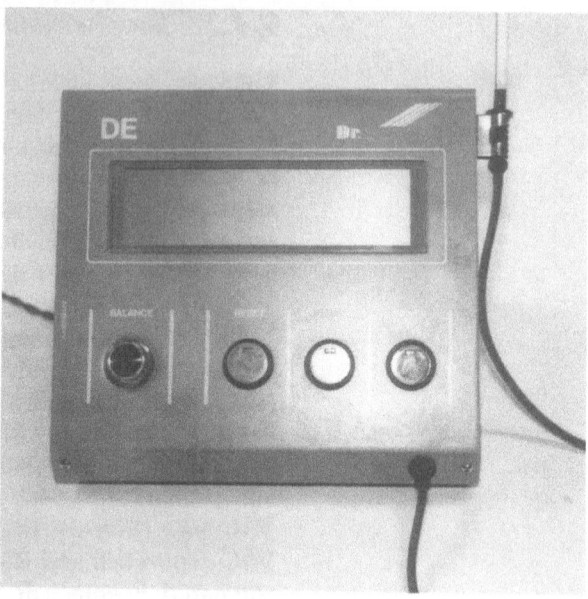

a   b

Fig. 9.2.a, b. Instruments for detecting and measuring halitosis. (Reproduced with permission from Winner's Japan Co., Ltd., Tokyo). Easy-to-use breath detector with a built-in gas sensor which acts as a substitute for an analyzer. The device can express the odor strength of the breath **b** A professional digital automatic mouth odor detector (DE 160) which is equipped with an ultra-sensitive semi-conductor gas sensor. When a gas is detected, value decreases are registered which indicate the degree of halitosis

selectively with high sensitivity. However, such measuring methods at laboratories are not available on a daily clinical basis requiring quick and simple examination.

In 1975, Tsunoda successfully detected hydrogen sulfide, methylmercaptan, and dimethyl sulfide in the breath of a patient afflicted with halitosis, and defined the corelation between them and the presence and degree of halitosis. Umezu (1975) succeeded in fixing the quantities of those volatile sulfuric compounds by using FPD and further confirmed the corelation between those substances and halitosis.

More recently, Tsunoda and Watanabe (1988) have developed two easy-to-use breath detectors (Fig. 9.2.a,b). These detectors have a specially developed breath sensor which is especially sensitive to volatile compounds such as methylmercaptan, and are used for measuring the degree of halitosis. The principle of these detectors is to send breath to the combustor equipped with a semiconductor which has a high surface temperature to measure the amount of methylmercaptan; as a result, the temperature of the semiconductor rises and the electrical resistance decreases. The difference in the electric charge of a pair of electrodes thus obtained is then converted to the odor density.

The variation of electric current in the semiconductor is calculated by microcomputer and the result indicated in a liquid crystal display. Methylmercaptan density is indicated in numerical value (ppm) and the halitosis level expressed in figures such as $-$ (less than 0.2 ppm), $\pm$ (0.21–0.3 ppm), $+$ (0.31–0.5 ppm), and $++$ (greater than 0.51 ppm) to indicate the degree of halitosis. Density is shown in a graphic chart to indicate halitosis at a low density that cannot be expressed in numerical value. This breath detector can easily be used at chairside, and it is also possible to indicate the result in objective numerical value. Thus, it can be very useful in diagnosis of patients afflicted with halitosis or in cases of psychogenic halitosis.

9.5 Foot Odor

Eccrine bromidrosis is a foot health problem which involves foot odor, sogginess, tenderness, and occasional blistering of the foot's interdigital areas and plantar weight-bearing areas. Commonly called "foot odor," this condition is due chiefly to microbial growth and skin metabolism. It is believed to occur due to increased moisture retention and absorption. Moisture that softens the skin may make it soggy, tender and less resistant to abrasion, thus providing a humid environment conducive to increased growth of microorganisms on skin surfaces. The toe webs, soles, and heels are the dampest foot areas; therefore, pressure and friction at these softened areas can produce blisters and infections. Bromidrosis may figure in serious infections even though it is chiefly a hygienic or social, rather than medical, problem (Abramson and Terleckyj 1979).

About 1,000 sweat glands are distributed over the area of 1 cm^2 of human sole skin, and approximately 16 ml of sweat is secreted in 24 h. Gram-negative bacilli and gram-positive cocci from the intestinal track are also more frequently encountered on the lower extremities, and these bacteria may colonize and become dominant on these skin sites, thereby contributing to foot odor. Increased foot hydration can stem from occlusive footwear and/or excessive sweating. The eccrine glands are the primary factor in foot secretions. Eccrine secretions are known to contain sodium chloride, vitamins, glucose, lactic acid, urea, potassium, fat molecules, and fatty acids in addition to a clear, hypo-

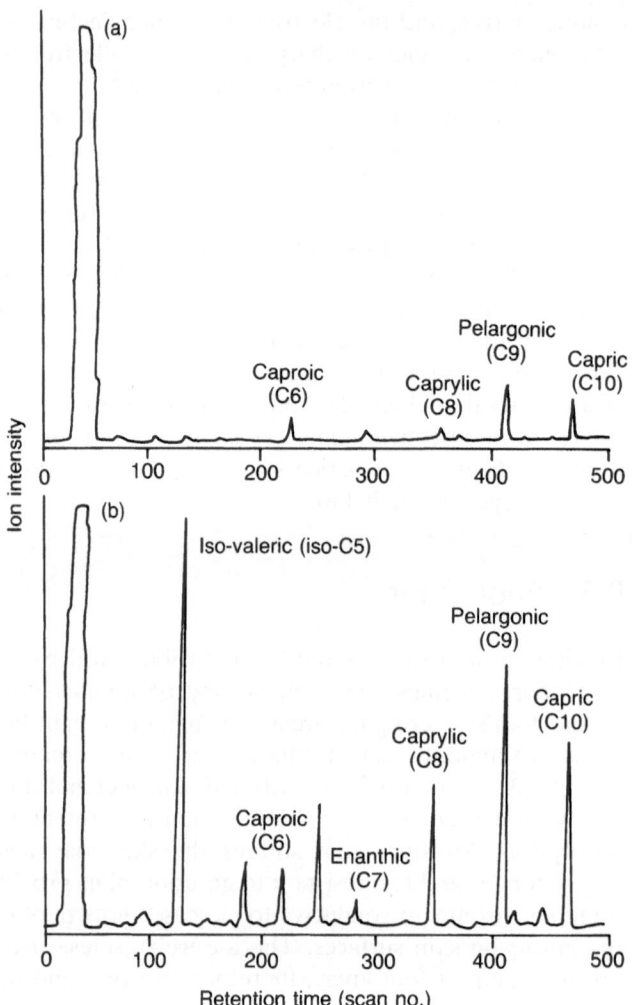

Fig. 9.3.a, b. GC/MS analysis of foot extract. Mass chromatograms from **a** a subject without foot odor and **b** a subject with foot odor. (Reproduced with permission from Kanda et al. 1990)

tonic, watery solution. Along with cellular debris, keratin, keratohyaline granules, and skin surface lipids, these substances serve as substrates for resident microflora and may be metabolized into volatile, odoriferous byproducts which give rise to quite a few of the characteristic foot odors.

Hyperhidrosis has major pathogenetic importance, but obesity, intertrigo, and diabetes mellitus also predispose, especially in cases of intertriginous bromidrosis. Diabetics show both hyperhidrosis and a predisposition to both bacterial and candidal infection, which contributes substantially to odor production.

Skin surface lipids partially degraded by gram-positive bacteria form unsaturated fatty acids which show strong antimicrobial activity against gram-negative bacteria and certaiñ fungi. Bacteria such as *Staphylococcus aureus* and *S. epidermidis*, *Corynebacterium acnes*, and other diphtheroids reportedly have lipolytic activity in vitro (Reisner and Puhvel 1969). The diphtheroids are chiefly implicated in foot odor due to their ability to metabolize complex lipids in sebum and epidermal lipids.

Kanda et al. (1990) reported that short-chain fatty acids from the socks and feet of subjects either with strong foot odor or with weak or no foot odor were extracted with ethyl ether, and then analyzed by gas chromatography/mass spectrometry (GC/MS). Short-chain fatty acids were found in greater amounts from those subjects with strong foot odor. Isovaleric acid was present in all the subjects with foot odor but was not detected in those without. Olfactory evaluations of the various short-chain fatty acid solutions were in agreement with the GC/MS analyses. By incubating sweat and lipid from subjects with strong foot odor, we succeeded in reproducing the foot malodor. GC/MS analyses of reproduced foot odor revealed that short-chain fatty acids were present in a similar composition to that found in vivo (Fig. 9.3.a,b).

9.6   Scalp Odor

The odor of unwashed hair may be unpleasant but is almost never a clinical problem. A peculiar mousy or musky odor described in cases of fungal infection apparently emanates from diseased scaly hair and skin (Hurley HJ Jr 1987). Labows et al. (1979a) reported an attempt to duplicate natural scalp odors in vitro by incubating normal skin microorganisms with sebaceous and apocrine gland secretions. The yeast *Pityrosporum ovale*, which is the major scalp resident, can metabolize lipid substrates to 4-hydroxy-acids which readily undergo ring closure to the volatile and odorous γ-lactones. The culture odor resembles that of unwashed hair and is a close match with that of γ-decalactone, the major lactone component. This scalp microorganism, due to the compounds it produces, could be used for natural formation of flavor additives. The scalp odor contains short-chain aliphatic acids as well as lactones. Formation of scalp odors may involve cooperation by *Propionibacterium acnes*, which is capable of hydrolyzing triglycerides, and *Pityrosporum ovale*, which is able to metabolize the resultant fatty acids and/or glycerol to various odorants.

Goetz et al. (1988) has analyzed hair and scalp odor in respect to short-chain fatty acids that have also been identified in body sites such as the axilla (Nitta and Ikai 1954), vagina (Preti and Huggins 1975) and foot (Kanda 1990).

9.7   Other Odors and Disorders

Hurley (1985) summarized the heritable metabolic disorders, in particular a group of these disorders in which distinctive intrinsic odors are detectable in sweat (Cone 1968; Scriver and Rosenberg 1973). The best known is phenylketonuria, a condition in which active phenylalanine hydroxylase is absent, resulting in hyperphenylalaninemia and increased phenylpyruvic acid and phenylacetic acid in sweat and urine, detected as a characteristic musty or sweaty "locker room" odor.

Maple syrup urine disease occurs in infants as a familial cerebral degenerative disorder characterized by defective oxidative decarboxylation of the branched-chain amino acids valine, leucine, and isoleucine. These acids and their α-keto analogs, which accumulate in urine, sweat, and cerumen, give rise to a malty, caramel, or maple syrup-like odor.

Another odor-producing disorder, called the oasthouse syndrome, derives its name from an odor given off by infants with this condition which resembles the odor of dried malt or hops in English oasthouses. It is a defect similar to phenylketonuria but with a different enzymatic block that leads to accumulation of α-hydroxybutyric and phenylpyruvic acids as well as methionine, phenylalanine, and tyrosine in sweat and urine. These infants, as well as those with phenylketonuria, undergo a change in hair color. The methionine malabsorption syndrome produces the same odor as the oasthouse syndrome, with accumulation of α-hydroxybutyric acid in sweat, urine, and feces (Hooft 1964). There is no phenylalanine metabolism abnormality, however, and a diet low in methionine will lead to gradual elimination of the symptoms.

Another syndrome, called hypermethioninemia, produces a strong fishy, fruity, sweaty, or rancid butter-like odor in sweat and urine. Infants afflicted with this condition reveal familial hepatic cirrhosis, renal tubular defects, and hypoglycemia. The odorant is suspected to be l-keto alpha-methiolbutyric acid. Another amino acid disorder, isovaleric acidemia, is characterized by an apparent deficit of isovaleryl coenzyme A (CoA)dehydrogenase which gives rise to very high levels of isovaleric acid in the serum and subsequently in the sweat and urine. The "cheesy" or "sweaty feet" odor described in infants with this condition is indicative of mental retardation and recurrent acidosis.

The N-butyric/N-caproic acidemia or "sweaty foot odor" syndrome is a rare disorder in which the affected infants appear to be normal at birth. However, by 7 days of age they develop weakness, lethargy, acidosis, and intense odor of the sweat, urine, and breath similar to that of sweaty feet (Sidbury et al. 1967). This condition is due to large amounts of butyric and hexanoic acids in the body as a result of enzyme acyl dehydrogenase deficiency.

Odors emitted by these patients should be almost diagnostic to the experienced clinician or at least point toward any such disorder. An important aspect is that the eccrine sweat glands—not the apocrine glands—seem to be the ordinary cutaneous excretory route of the odorants in these patients, who are usually so young that the apocrine sweat glands are not even functional. The sole exception seems to be maple syrup urine disease, a condition in which the odorous chemicals are also excreted in the ceruminous or wax glands of the external auditory meatus. These apocrine-type glands are fully functional in infancy, whereas the apocrine glands of the axilla and other skin areas are not. If these patients survive and can be studied up to and beyond puberty, their apocrine sweat might be similarly odorous.

Another odor-producing disorder found in adults as well as children is trimethylaminuria (Lee et al. 1976). If they consume large quantities of fish, individuals with this condition cannot completely degrade the digested fish. This results in high levels of di- and trimethylamine, and gives rise to a "fishy" odor in the breath and eccrine sweat. The deficiency of hepatic trimethylamine oxidase, which is the metabolic defect in these patients, seems to be inherited as an autosomal dominant trait. It is not clear if the odorogen is excreted in the apocrine sweat of these patients.

9.8   Feces Odor

Among the patients who are annoyed by body odor, there are those who persistently complain that their bodies smell of feces. Abiko (1986) defines the source of feces and flatus odors as the gasified composition of dissolved products of residual food, produced by the action of intestinal bacteria existing in the lower small intestine and the large intestine. In the case of a healthy person, the food residue in those areas is composed mostly of food fiber, and the gasified composition of the dissolved products produced by bacterial action is composed mostly of low fatty acid. This composition emits a mild and sweet odor. When dissolved products other than those of food fiber are added to it, feces and flatus produce offensive odors. If the food residue is composed mainly of glucide, it is called "fermentative" and has a sour odor. On the other hand, when it is composed mainly of protein, it is putrid and emits an offensive odor represented by hydrogen sulfide gas.

Chapter 10.   Body Odor as a Psychosomatic Disorder

Secretions of apocrine and eccrine glands can be stimulated by psychosomatic factors. Sweating is not an exclusively physiological phenomenon. It can be stimulated by a nervous condition stemming from concern over sweating.

Both mental and physical factors in combination can frequently cause illness of a psychosomatic type. For example, stomach ulcers are generally thought to be a physical disorder. However, ulceration can typically result from a mental disorder. A nervous individual is more likely to develop an ulcer than someone who is characteristically calm. Mental suffering can cause stomachache and loss of appetite; stomach acid may then eat through the stomach lining to produce the perforation of an ulcer.

Stress, which is caused by mental suffering, affects the autonomic nervous system. Recognition of a stomach problem caused by stress can aggravate the condition. That is why medicines which block the operation of the autonomic nervous system are used for treatment of ulceration.

Individuals who sweat excessively (hyperhidrosis) can also suffer from psychosomatic disorders (PSD). Since the intensity of body odor can be influenced by sweat volume, body odor patients frequently reveal psychosomatic disorders that exacerbate the problem and may create a body odor neurosis or olfactory paranoia.

Until Kushima (1966) discussed PSD at an academic congress in Japan held in 1949, psychosomatic disorders were treated only as forms of neurosis, but the combined mental and physical symptoms are now commonly recognized.

Hasegawa (1971) has singled out three characteristics of PSD: (1) an obvious psychological factor is involved in the onset of the disorder, (2) even if the illness can be basically ascribed to physical causation, the patient's negative mental condition and character adds complications, and (3) neurosis can include actual physical illness.

Thus apocrine and eccrine gland secretions can be influenced by psychological factors: the more the individual is distressed by sweating, the more he will tend to sweat, producing a vicious cycle. A typical case that exhibits this pattern of organic disorder is patient A in Fig. 10.1. On the other hand, when the physical factor is not as obvious as the psychological factor, the case is called a psychosomatic disorder (patient B). When the psychological factor far outweighs the physical factor, it is called an organic neurosis (patient C). If the symptoms of patient C take a turn for the worse, the condition must be classified as a mental disorder (patient D). When this disorder reaches the point that the patient experiences delusions or begins to impugn his own identity, the disorder becomes profound, and the patient may have to be institutionalized (patient E).

In the diagram, F stands for psychosomatic disease on the whole, without A and E classifications. However, in respect to cases A through D, lines should not be drawn too clearcut, because distinctions are often quite subtle. Moreover, all of these conditions are curable. It is important to consider the patient's physical and psychological profiles in terms of efficacious treatment procedures.

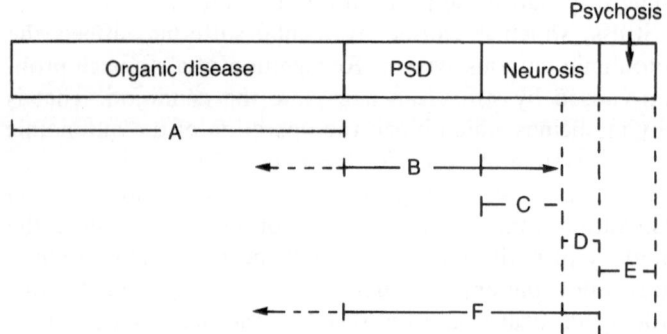

Fig. 10.1. Relationship between bromidrosis and neurosis. Patient *A* exhibits a pattern of organic disorder; patient *B*, psychosomatic disorder (*PSD*); patient *C*, organic neurosis; patient *D*, mental disorder; patient *E*, psychosis. We have added *F* to indicate a wide spread of PSD. (Reproduced with permission from Inaba 1986)

Table 10.1. The Shimada neurotic test examines the degree to which patients are nervous. It consists of three parts: *H*, health, *S*, sociability, and *E*, emotion. Patients were requested to answer a total of 50 questions (13 for H, 12 for S, and 25 for E) and the averaged results are shown under each column. The figures indicate that the patients before the operation are generally much the same as normal persons in these three categories. Both emotion (E) and health (H) indicate less nervousness after the operation and an increase in sociability (S). (Reproduced with permission from Inaba 1986)

	No. of cases	*H*	*S*	*E*	*Total*
Patient before bromidrosis operation	140	2.84	1.53	6.99	11.36
(normal person)		(3.00)	(1.81)	(7.32)	(12.13)
Patient after bromidrosis operation	120	2.58	2.50	4.57	9.64

10.1 Examination of Psychosomatic Disorder (PSD) Patients for Psychological and Social Adjustment

There are three well-known tests used in Japan to determine a patient's state of mind. Identifying that state of mind can be quite important, since psychological factors influence the degree of illness.

10.1.1 Shimada Neurotic Test

This test (Shimada 1953) examines the degree to which patients are nervous. It consists of three parts, H (health), S (sociability), and E (emotion), and each part contains 50 questions. As Table 10.1 indicates, it is not true that bromidrosis patients are typically more nervous than normal people. State of emotion (E) before surgery (6.99) is lowered after the operation (4.57). The feeling of sociability (S) is somewhat improved. Very little difference is observed in the basic feeling of health (H) before and after the operation. After the operation, the patient calms down and feels relieved.

10.1.2 Yatabe-Guilford (Y-G) Test

The Y-G (Yatabe-Guilford) test examines the personal character of the patient. It consists of three parts: emotional state, social adaptability, and character tendency. From

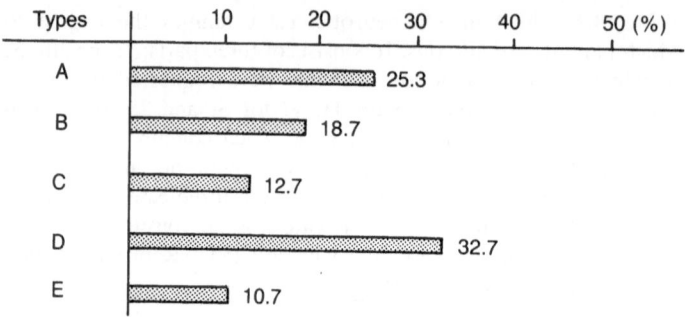

Fig. 10.2. Character of bromidrosis patients by the Y-G test (Yatabe-Guilford test). Type B (18.7%) and E (10.7%) patients who suffer from body odor show emotional instability and a lack of social adaptability. (Reproduced with permission from Inaba 1986)

these three categories, we can detect roughly five kinds of character types. Type A is the average type. Type B is emotionally unstable, lacks social adaptability, and is extroverted. This type is the quasi-neurotic type. Type C is very inactive and introverted. Type D is emotionally unstable and lacks social adaptability, but is active; this type rarely shows the characteristics of hysteria. Type E is emotionally unstable, lacks adaptability, and is introverted. This type is typical of neurosis and is often found among neurosis and PSD patients. In our results, 25.3% of the patients were type A. About another 30% who suffer from body odor were categorized as types B (18.7%) and E (10.7%), showing emotional instability and lack of social adaptability (Fig. 10.2).

10.1.3   Cornell Medical Index (CMI) Test

A lengthy all-inclusive and self-administered personal history form developed at the Cornell University Medical

Table 10.2. Cornell Medical Index (CMI) test

Type	Condition	Female	Male	Total Subjects	%
I Primary	Normal	36	11	47	42
II Secondary	Almost normal	37	5	42	38
III Third	Somewhat neurotic	16	1	17	15
IV Fourth	Neurosis	5	0	5	5
	Total	94	17	111	100

(Reproduced with permission from Inaba 1986)

School, this test contains 195 questions and is categorized into four different types. It indicates that patients should be treated in the way that matches each patient's character most closely. In aggregate of the above findings, some bromidrosis patients are shown to react neurotically to the odor, a factor which restrains their personal freedom of action and pressures the patient to take on a different form of personal character (Table 10.2).

10.2 Physical Examination of Body Odor

Since the autonomic nervous system controls sweating, we have tests which examine whether the system is in balance and functioning correctly. The balance is, of course, fundamentally controlled by the brain. These exams are:

1. Medical examination (degree of epinephrine test, amount (%) of porphyrin, amine, and catechol detected in blood and urine)
2. Vascular system examination (Aschner's phenomenon, etc.)
3. Respiration system examination (respiration curve)
4. Dermovascular reaction (dermography)
5. Electrical examination (galvanic skin reflex, microvibration, etc.)

Even at present, it is quite difficult to determine whether the autonomic balance system (ABS) is functioning correctly by only one result. When a female patient goes into menopause, it is usually a sign of ABS loss. If ABS control is lost, there is a high possibility of body odor arising from uncontrollable sweating.

Patients who have body odor are not necessarily PSD patients, but this is true most of the time. Those patients whose physical conditions produce body odor but who do not suffer psychologically from the odor are not in the PSD category but have a purely organic disorder. In the Y-G test, about 30% of all patients belonged to the unstable and incongruent type. In the CMI test, about 20% had a tendency toward neurosis. Approximately 20%–30% of all patients are thus considered to be of the PSD type.

There are other factors which might affect PSD patients:

1. Autonomic nervous disorder (abnormal sweating in puberty and menopause due to hormone imbalance)
2. Genetic-superior genes are dominant as a heredity factor, giving rise to psychosomatic stress

3. Sense of smell (abnormal sense of smell, hyperosmia, toward one's own body odor and overconcern that body odor may be detectable to others even if the subject cannot detect it—on the contrary, if the patient has hyposmia, he or she has recognized his or her own body odor by reference to other people's behavior)

Some people are more likely than others to become neurotic. They include: (a) patients who lack self-confidence, (b) perfectionists (obsessional neurosis—due to an excessively fastidious character, the patient wants perfection in everything), (c) depressive, and (d) selfish (hysteria—this is a condition that presents somatic symptoms, simulating almost every type of physical disease, along with a series of mental symptoms. If the symptoms are serious, it is probable that the patient will prove to be neurotic).

Chapter 11. Body Odor as Delusion

11.1 Case Reports

A certain number of patients who complain of body odor exhibit perplexing psychological problems. They become obsessed with alleged body odor with no objective evidence to support their concern. Many of them have consulted other people who could not verify their symptoms, and yet these patients persist in the belief that their body odor is offensive. Eventually, most of them withdraw from the care of their usual physician, only to start a search for remedial treatment anew. Many of these patients have visited several different clinics and hospitals seeking permanent relief from this presumptive malodor. Since there has been no completely effective treatment in the past, especially for bromidrosis, they were not fully satisfied with any treatment received. On the other hand, even when effective treatment is completed, they refuse to accept the fact, persisting in the belief that nothing can rid them of offensive body odor. Some of these patients indicate symptoms as described below and in Table 11.1.

Case 1. Female, 51 years old, wet cerumen. From the age of 30 years, she claimed to have a foul body odor. She was treated by electrocoagulation several times, but the condition persisted. She was then treated three times by radical removal surgery, with a large scar (5 × 8 cm) remaining and no axillary hair visible. Objective evaluations failed to find any odor, but the patient still claimed that, in spite of many remedial procedures, the malodor still remained, as evidenced by people in her immediate vicinity showing unpleasant facial expressions, touching and wrinkling their noses, or opening a window. In other words, she interpreted the behavior of people around her as proof of

Table 11.1. Summary of case reports of odor as delusion

Case	Sex-Age	Symptoms developed at age	Cerumen type: + dry − wet	Treatment methods	Objective evaluations		Subjective evaluations			Various tests			Treatment methods	Results
					Axillary hair	Axillary odor	Axillary odor	Axillary hyperhidrosis	Hyperhidrosis in general	Degree of odor sensitivity	Character by Y-G Test	Severity of neurosis		
1	♀51	34	−	Electro-coagulation ×2	−	−	+++	−	−	4.66	B	20	Subcutaneous tissue shaving method ×2	Worsened
2	♀19	17	+		±	−	+++	++	++	4.66	E	15	Subcutaneous tissue shaving method	Slightly improved
3	♀47	27	−	Electrolysis ×5	−	−	++	++	++	6	E	8	Psycho-therapeutics	Slightly improved
4	♀50	47	−	Removal method	−	−	++	−	++	4.66	AE	14	Subcutaneous tissue shaving method	No change
5	♀45	22	−	Electrolysis ×2 Removal	−	−	++	+	++	8	A	14	Subcutaneous tissue shaving method	Slightly improved
6	♀35	20	+		+	−	+	+++	+++	8.44	B	21	Subcutaneous tissue shaving method	Totally improved
7	♀47	42	−	Electrolysis Removal	−	−	++	±	++	3	E	20	Subcutaneous tissue shaving method	Totally improved

8	♂26	20	−	Electro-coagulation	−	−	++	+++	+++	Lost	E	20	Subcutaneous tissue shaving method	Totally improved
9	♀32	20	+		++	−	++	++	+	4.66	AE	19	Subcutaneous tissue shaving method	Slightly improved
10	♀33	20	−	Removal	−	−	++	+	+	4	E	14	Psycho-therapeutics	Unknown
11	♀40	39	−	Electrolysis ×5	+	−	++	++	++	5.66	A	17	Subcutaneous tissue shaving method	No change
12	♀38	Unknown	−		−	−	+	±	+	4	C	18	Psycho-therapeutics	Slightly improved
13	♀22	18	+		+	−	+	±	+	4.66	AE	19	Subcutaneous tissue shaving method	Unknown
14	♀36	20	−		−	−	++	−	−	2.66	B	14	Subcutaneous tissue shaving method	Unknown
15	♂20	19	−	Removal	±	−	++	±	+	4	C	14	Subcutaneous tissue shaving method	No change

(Reproduced with permission from Inaba 1986)

her repugnant body odor. The authors performed the sub-cutaneous tissue shaving procedure on this patient after detecting hyperhidrosis due to remaining eccrine glands in surrounding axillary scar tissue, but she insisted that the odor persisted after treatment. Examination results revealed her degree of odor sensibility as 4.66, the Y-G test, type B, and the severity of neurosis (Shimada test) was 20 (health 6, sociability 3, emotional condition 11).

Case 2. Female, 19 years old, dry cerumen. This patient did not have bromidrosis, only hyperhidrosis. However, she was convinced she had a foul odor. Objectively, we could not detect any odor. She claimed that when friends came close to her, they smoked cigarettes, and many people pass-ing her wrinkled their noses. "They don't want to get close to me," she complained. She worried constantly about inconveniencing other people; she could not concentrate on her work and usually perspired in the armpits. Her odor sensitivity was tested at 4.66, the Y-G test was type E, and the severity of neurosis was 15. Convinced that psy-chosomatic sweating caused the malodor, she agreed to subcutaneous shaving surgery and apparently gained new confidence. After 1 year she returned to our clinic and reported that she did not worry about those around her so much and was no longer self-conscious about possible body odor.

Case 3. Female, 47 years old, wet cerumen. This patient was anxious about axillary odor and had gone through five electrocoagulation sessions before undergoing our radical treatment procedure. After 1 year had passed, however, she came back and said, "I have vaginal malodor and want my womb removed." We refused to perform any such operation, telling her that the odor would not diminish even if her womb were removed. One year later she came back again insisting that she still had axillary malodor and wanting a reoperation performed. Her test results showed degree of odor sensitivity at 6, indicating slight hyposmia, the Y-G Test was type E, and the severity of neurosis was 8. This patient was convinced that body odor was emitted with sweating, especially during her current menopause. Even after axillary surgery, she still insisted that the odor persisted. Female patients may believe that foul odor emanates not only from the axillae, but also from a per-foration of the womb that permits gas to escape. These patients incessantly seek remedial treatment from a variety of physicians, one after another.

Case 4. Female, 50 years old, wet cerumen. The patient was treated by radical removal surgery. The degree of sensitivity was 4.66; Y-G test, type AE; and severity of neurosis, 14.

Case 5. Female, 45 years old, wet cerumen. The patient had gone through two electrocoagulation sessions and radical removal surgery. She was anxious about body odor due to hyperhidrosis caused by an autonomic nervous system disturbance (menopausal disorder). Tests indicated hyperosmia (8); Y-G test, type A; and severity of neurosis, 14.

Case 6. Female, 35 years old, dry cerumen. The patient was anxious about hereditary disposition. Tests indicated that she bore a grudge against her parents and had not married. She indicated hyperosmia (8.44); Y-G test, type B; and severity of neurosis, 21.

Case 7. Female, 47 years old, wet cerumen. Patient has hyperhidrosis due to a menopausal disorder. The degree of sense of smell was hyperosmia (3); Y-G test, type E; and severity of neurosis, 20.

Case 8. Female, 26 years old, wet cerumen. The patient worried about a hereditary factor and hyperhidrosis in spite of receiving electrocoagulation treatment. Anosmia was indicated, and the Y-G test result was type E. The severity of neurosis was 20.

Case 9. Female, 32 years old, dry cerumen. The patient has anthropophobia. She had worried about her own body odor, avoiding contact with other people. The degree of sense of smell was normal; Y-G test, type AE; and severity of neurosis, 19.

Other cases omitted from this discussion are shown in Table 11.1.

11.2 Psychological Diagnosis

We discussed the histories of these patients with a clinical psychiatrist whose diagnosis was as follows: (1) depression due to menopause in cases 1, 4, and 7; (2) neurosis in case 6; (3) schizophrenia in case 3 and (4) depression in case 2. The authors were surprised by the difference in opinion between the psychiatrist and a PSD consultant. We had attributed these disorders to PSD. In specific cases, however, we recommended that the patients consult a psychotherapist.

11.3   Clinical Characteristics

Cases of this kind account for about 15 out of 1,000 patients. By sex, females outnumber males by a ratio of 13:2. In terms of wet or dry cerumen, the ratio is 12:3, respectively. In reviewing previous treatment, six cases reported experience with electrolysis and another six had undergone some type of radical surgical treatment. Almost all of them claimed to have hyperhidrosis in spite of these procedures. Some stated that the condition persists in tissues surrounding the scar surface in the cases of surgical treatment. Those cases treated by electrolysis claim persistent hyperhidrosis because the eccrine glands are independent of the hair follicles and have not been removed. All of these patients claim that sweating occurs over the entire body surface, especially during puberty and menopause. This condition may be related to an autonomic nervous system imbalance.

The syndrome may also be closely related to olfactory sensitivity. Among the pertinent 15 cases, 9 had normal olfactory sensitivity, another 3 had hyperosmia, and the remaining 3 had hyposmia. The personal character of these patients, indicated by the results of Y-G testing, shows 20% in group B, 33.3% in group E, 20% in group AE, and a total of 73.3% in groups B, AE, and E (Fig. 11.1). In other words, 73.3% indicate emotional instability. This rate is higher among bromidrosis patients (group B 18.7%, group E 10.7%, for a total of 29.3%).

In the Shimada neurotic test for nervousness (Table 11.2) 4.1% showed a healthy tendency, 3.2% indicated sociability, and 8.8% indicated emotional problems, for a total of 16.1%. This tendency is higher in PSD patients than in both normal control subjects and bromidrosis patients. This indicates higher levels of neuroticism (emotional instability levels). Because these patients are hypersensitive about

Table 11.2. Comparison of psychosomatic and general bromidrosis patients by the Shimada neurotic test

	Healthy (H)	Sociable (S)	Emotional (E)	Total
Psychosomatic bromidrosis (PSD)	4.1	3.2	8.8	16.1
General bromidrosis	2.8	1.5	6.9	11.4
Normal	3.0	1.8	7.3	12.1

Tendencies of H, S, and E are seen to be higher in PSD patients than in normal subjects and bromidrosis patients. (Reproduced with permission from Inaba 1986)

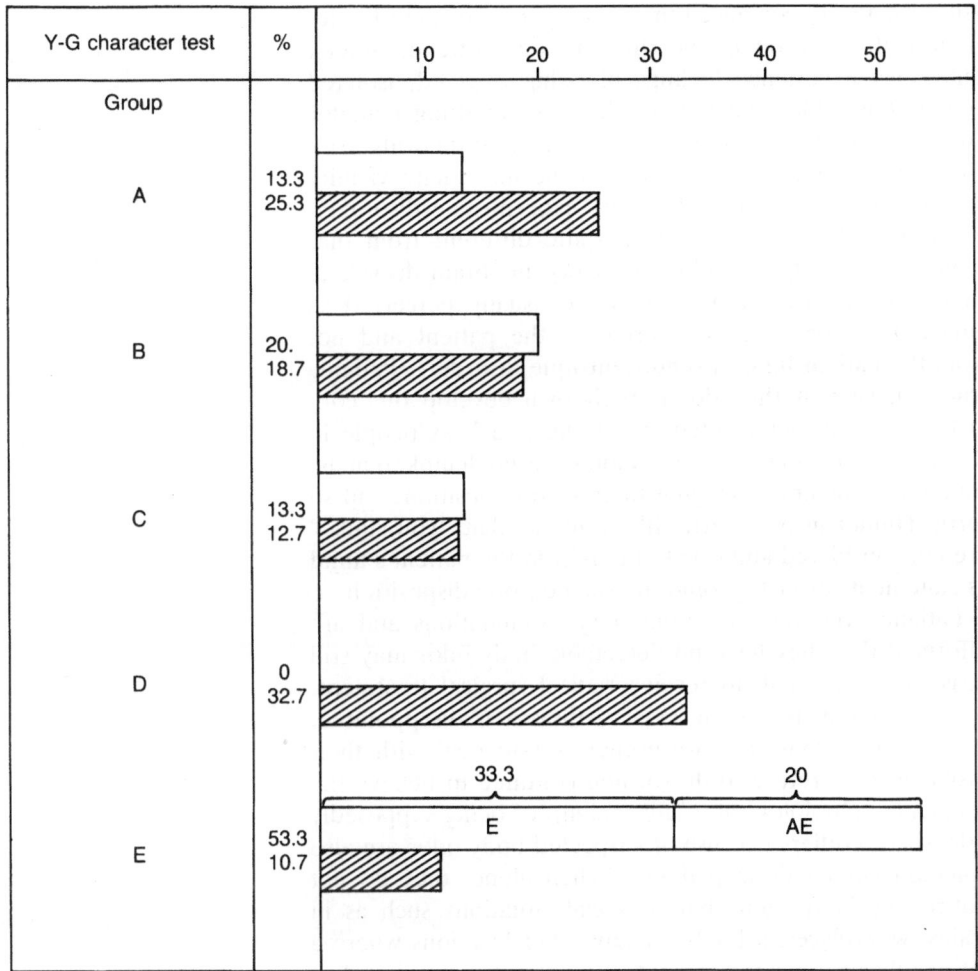

Psychosomatic bromidrosis

General bromidrosis

Fig. 11.1. Comparison of the personal characteristics of psychosomatic bromidrosis patients and general bromidrosis patients by Y-G testing. For general bromidrosis patients, the results shown in Fig. 10.2. were used. (Reproduced with permission from Inaba 1986)

their own body odor even though normal people can detect no odor, they persist in believing that other people do smell their sweat odor. These patients show indications of extreme self-consciousness or olfactory paranoia.

Kasahara et al. (1972) reported on characteristic cases.

The patients are convinced that some part(s) of their bodies emit(s) offensive odor, and that this odor offends others. They may experience feelings of being rejected, isolated and contemptible, and are convinced of emitting malodor due to a physical disorder. They remain stubbornly convinced that whatever they smell in the immediate vicinity comes from their own body odor (exocentrique disorder). Their mental state, however, is quite different from that found in cases of schizophrenia or organic brain disorders. In the latter cases, those odors are usually perceived as emanating from sources external to the patient and not from the patient himself (endocentrique disorder). Patients who believe that the odor is their own develop this conviction from misinterpreted social cues, such as people in the vicinity opening doors or windows, using handkerchiefs, sneezing, coughing, changing their seating position, and so forth. Humor in poor taste, like remarks that "you stink," are not considered amusing but persist in the patient's mind as statements of fact, promoting the neurotic disposition.

Patients who receive preliminary examinations and are informed that they have no detectable body odor may still insist that they wish to receive radical surgical treatment. If assured this is not necessary, they feel disappointed. If advised to consult a psychotherapist to deal with their problem, they refuse to do so, and continue to believe the problem is physiological. Guilt feelings of being supposedly offensive to others because of suspected body odor are also common among these patients. When alone, they do not notice any body odor, but in social situations such as in trains, workplaces, schools, or any other locations where a lot of other people are present, the patient is convinced of emitting an odor which is offensive to others. This is due purely to misinterpretation of other people's attitudes and actions. Given this situation, the patient will prefer to be alone but will also have a desire to seek effective treatment. Since, however, body odor is not the problem but only the mistaken fear of it, the required treatment lies in psychotherapy.

Body odor does not necessarily emanate from the whole body but only from some part(s), such as the armpits or pubic zone. In other words, because the locale of origin is strictly limited, it is easy to relieve. If it is relieved, patients should experience immediate psychological relief. Patients who claim that the odor persists frequently demonstrate a lack of belief in statements to the contrary made by doctors and nurses. To counteract this, the authors introduced an alternative practical testing procedure. Patients are invited to smell each other after testing their own degree of olfactory sensitivity, and thus gain confidence in discovering that the partner reports no detectable body odor. A psychotherapist then consults the patient about individual body odor and provides behavioral counseling. This approach has shown good results.

11.4   Identification of Olfactory Paranoia

As stated earlier, there is a close relationship between the condition of cerumen and bromidrosis: patients with dry cerumen have almost no odor. Accordingly, even if patients with dry cerumen claim to have bromidrosis, the condition is almost always essential hyperhidrosis, not bromidrosis. Apart from those patients who indicate anosmia, almost all patients can recognize their own body odor, especially the difference before and after taking a bath. Given that general fact, any finding that a patient cannot recognize his or her own body odor indicates that no detectable body odor is present. For that reason, patients should be tested for olfactory sensitivity.

Another indicator is clothing discoloration in the axillary region. Bromidrosis patients do show this discoloration, since apocrine sweat includes fatty acids, cholesterols, and pigments. If discoloration is not observed, the sweat most likely contains no perceptible odor. People are usually reluctant to bring body odor to the offender's attention. However, individuals who in fact have offensive body odor can often recognize it through the behavior of the people around them. However, when another's behavior is misinterpreted to fit the patient's own preconception, a delusion of persecution may develop. Patients are well-advised to question family members and close relatives if they fear offensive body odor, rather than attempting to educe honest answers from friends and acquaintances who may be reluctant to discuss the matter. Even if the patient has received treatment for complete subcutaneous tissue removal in the axillary region with no regrowth of axillary hair, hyperhidrosis—but not bromidrosis—may still be observed. For patients who show considerable regrowth of axillary hair following other types of treatment with concomitant persistence of bromidrosis, we recommend surgical treatment (clearance method) for complete removal of hair follicles including eccrine and apocrine glands.

Janet (1903) classified the psychological profiles of patients displaying olfactory delusion as obsessive neurosis. Durand

(1955) ascribed it to melancholic delusion. Walter (1965) described it as "phobische Beziehungs syndrom" (dysmorphophobia). He explained the condition as one of the phobias concerning personal appearance: patients claim to have the odor only in social settings, not in their own homes. Adachi (1961) described this disorder in connection with specific patients, pointing it out as a disorder which has drawn interest as a specific syndrome different from psychosis. Shikano (1960) calls it a type of "chronic hallucinative disorder" and claims it does not involve schizophrenia or neurosis. In general, it is a problem of the patient's own self-consciousness.

Kasahara et al. (1972) depicted it as a syndrome with indications ranging from neurosis to schizophrenia, but said that, as a neurosis, it cannot develop into true schizophrenia but will tend to decrease in severity toward the end of the patients' 20s or 30s. Miyamoto (1976) calls it a disorder similar to olfactory paranoia. Bishop (1980) terms it "monosymptomatic hypochondriacal psychosis." Patients who exhibit this syndrome manifest specific systematic delusions with signs and symptoms of a corollary psychological disorder or impairment. Three subtypes of this syndrome have been described. With the first, called dysmorphophobia, patients show concern and conviction that body parts, in particular the face, are somehow disfigured, despite objective evidence to the contrary. The second subtype involves delusions of infestation, or parasitosis: patients believe, and behave as if they are infested with parasites or insects despite negative medical findings. The third subtype is that of olfactory reference syndrome or autodysosmophobia. In this subtype, delusions of repugnant body odor may be classified as a variant.

Chapter 12. Treatments for Palmar, Foot, and Genital Hyperhidrosis and Bromidrosis

12.1 Treatment of Palmar Hyperhidrosis

Body odors are not related to palmar hyperhidrosis, but hyperhidrosis caused by bromidrosis is closely related to it. Overconsciousness of body odors is likely to cause palmar hyperhidrosis. Referral to a dermatologic clinic has usually been followed by treatment with topical antiperspirant preparations and anticholinergic drugs. The side effects of the latter are undesirable, however, and it is unusual to find that such agents have brought about even a temporary amelioration of symptoms (Hartfall and Jochimsen 1972). Among all tested remedies for palmar hyperhidrosis, iontophoresis seems to be the most recommendable (Levit 1980). Some studies have shown that adding an ionizable material to the water is not requisite to obtain a therapeutic effect. Simple tap water is sufficient (Shrivastava and Singh 1977; Levit 1980; Stolman 1987). Good results are obtained if treatment is performed two or three times a week over a period of 3 weeks, followed by maintenance treatment once a month. A radio frequency technique used to destroy sympathetic ganglia seems to be effective in treatment for palmar hyperhidrosis (Wilkinson 1984).

Alcohol block of the cervicodorsal sympathetic ganglia has been advocated by Levine and Harris (1955). This is unsatisfactory in that the effect is unpredictable and usually not permanent. The procedure has been reported to cause troublesome intercostal neuralgia for many years afterward. Another recognized disadvantage of chemical sympathectomy is the concomitant development of Horner's syndrome, because the ciliospinal center is not sharply confined to the Th_1 spinal level, but may extend downwards as low as Th_5 (Baker and Baker 1975, Adar et al. 1977) (Fig. 12.1).

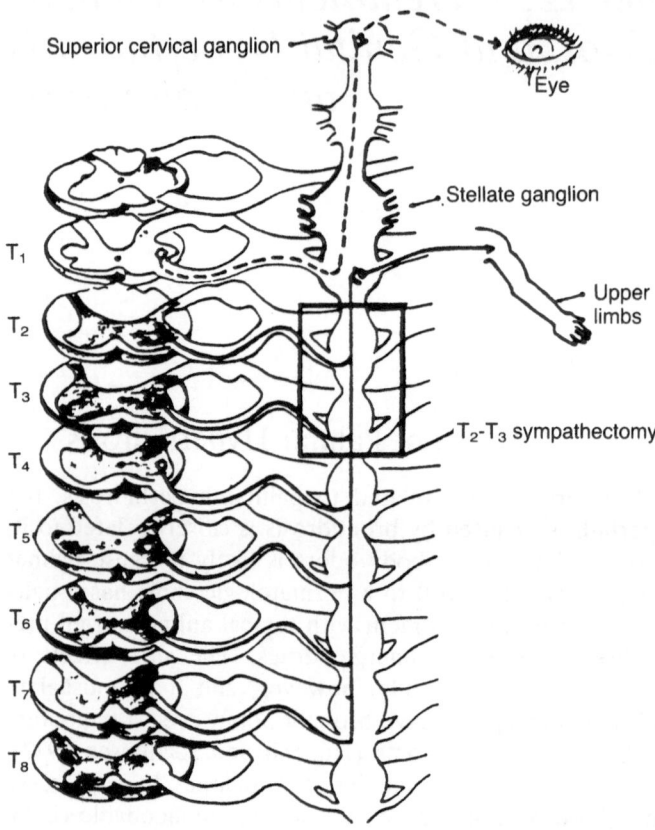

Fig. 12.1. Schematic diagram of the cervicodorsal sympathetic ganglia. (Reproduced with permission from Adar et al. 1977)

Since medical therapy for these conditions is often in-effective, a sympathectomy may be provided to the patient with severe essential or axillary hyperhidrosis (Adar et al. 1977; Welch and Geary 1984), but the results are inconsistent, most likely due to the variation observed in anatomic pathways (Greenhalgh et al. 1971). The success rate of Th_{2-3} sympathectomy has been reported as high as 92%–99% (Dohn and Sava 1978; Shih and Wang 1978). But as Dohn and others have reported, the complications of sympathectomy include compensatory hyperhidrosis (increased sweating at some other part of the body) in 24%–44% of all cases (Dohn and Sava 1978; Cloward 1969), pneumothorax in 10%–15% (Cloward 1969; Dohn and Sava 1978), permanent Horner's syndrome in 0.8%–4% (Cloward 1969; Dohn and Sava 1978), wound infection in 0.2%–2% (Dohn and Sava 1978; Ellis 1975), hemothorax in 0.2% (Dohn and

Sava 1978), intercostal neuralgia in 5% (Dohn and Sava 1978), and empyema in 1.5% (Dohn and Sava 1978). Sympathectomy is seen as efficacious for palmar hyperhidrosis, but the problems attendant on this form of surgical intervention are not negligible.

12.2 Treatment of Halitosis

Two methods that can be applied in the effective control of malodors of the mouth are: (a) removal of the periodontal problem through proper treatment (curettage, surgery, tongue brushing, rinsing with zinc ion), and (b) neutralization or masking of the odors. A 75% reduction of the odor-producing components can be accomplished by cleaning the dorsoposterior part of the tongue and rinsing with a mouthwash containing zinc. Brushing is only 25% effective. The combined brushing and zinc wash prior to bedtime reduced the volatile sulfur compounds of early morning mouth breath by approximately up to 10 h after treatment (Claycomb and Schearer 1986).

12.3 Specific Treatments for Foot Odor

Increased activity followed by moisture retention and absorption is not healthy. Fatigue and physical stress which contribute to foot moisture and odor are involved not only in bromidrosis but in other foot pathology.

Increased foot moisture can be attributed to poor footwear as well as excessive sweating. With respect to footwear, the dispersion of sweat can be impeded by compression of the interdigital region and by restriction of movement in the various parts of the foot, a condition which can be aggravated by a poor choice of footwear. Polyvinyl chloride shoes or those with polyurethane uppers are clearly more confining than those made with natural or artificial leather. Socks are also known to differ in moisture absorption according to their material.

Foot malodor can first of all be reduced by rigorous hygiene. A different pair of shoes should be worn each day with clean socks, even if the socks must be changed three times a day. Socks made of cotton and hemp absorb foot moisture better than nylon. A 10% boracic acid powder on the foot and in the shoe used daily is effective in diminishing the offensive odor. For persistent cases, X rays may be utilized, which will result in a decrease of the glands.

Also psychotherapy may be tried in severe cases, while sympathectomy produces complete and lasting dryness of the skin (Hauser 1945).

Cutaneous infections which include athlete's foot may be latent as dry, scaly lesions, but with increased foot moisture result in the wet lesions characteristic of athlete's foot caused by the flourishing growth of microflora which reside on a humid foot. Topical treatments are available to relieve these pathologic conditions.

12.4   Treatment of Genital Bromidrosis

There is a low rate of genital bromidrosis in the Japanese population (Inaba et al. 1973). Much higher rates are present in Caucasian populations. We have examined many patients who complain of malodor in the genital region, but it is rarely as serious as they claim. Even if there is a slight malodor and clothing discoloration, the condition is natural in female patients due to normal vaginal discharge.

Almost all of these patients are found to be excessively self-conscious or subject to olfactory paranoia. Deodorants used in the axillary region may be too strong for the sensitive vaginal mucosa, so mild deodorants are best recommended.

Complete surgical relief of genital bromidrosis is difficult because of the wide area which takes in the pubic and vulvar regions. The surgical method we use is much the same as the axillary shaving method. A 1-cm incision is made in both inguinal regions, followed by dissection with scissoring. Subcutaneous shaving is performed in the pubic vulvar areas as in the axillary region. The shaved skin is pressed down by the double tie-over dressing. However, since almost all of the dressing in the vulvar region is contaminated by urine secretion, we prefer a total resection of the hair-bearing vulvar region.

Malodor due to vaginal discharge is different and must be treated by a gynecological specialist.

Chapter 13. Pharmaceutical and Physical Treatment Procedures for Axillary Bromidrosis

Treatment procedures for relief of bromidrosis have been less than exact. Although medical knowledge has improved, physicians remain perplexed as to efficacious procedures for treating this disorder. Bromidrosis is not, of course, a dangerous type of disorder. In Western countries it is considered integral to a normal physiological constitution. Physicians in Japan also consider it a normal phenomenon and see no need for radical treatment. They cannot understand why bromidrosis patients are so concerned about slight odor and tend to refer them to neurologists for treatment as olfactory paranoids. As stated, however, these patients are not concerned only about the odor but the hyperhidrosis and concomitant clothing discoloration as well. Pharmaceutical and physical treatment procedures have consisted of:

1. Pharmaceutical therapy (topical antiperspirants)
   a) Local pharmaceutical therapy
   b) Systemic medication therapy
   c) Local injection therapy
2. Physical therapy
   a) Electrolysis and electrocoagulation
   b) Iontophoresis
   c) Radiation therapy (X-ray therapy, radium, etc.)

13.1 Pharmaceutical Treatment

13.1.1 Local Pharmaceutical Therapy

Local pharmaceutical treatment has included use of odor suppressants such as topically applied aluminum chloride, potassium permanganate, formaldehyde solution, zirconium, zinc deodorant preparation, and glutaraldehyde or oral

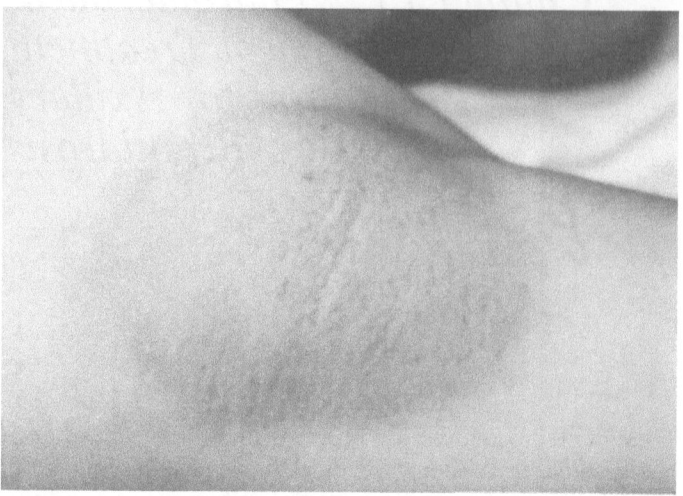

Fig. 13.1. Contact dermatitis caused by deodorant application. Frequent application and strong concentration of active ingredients may irritate the skin

anticholinergic compounds. However, none of these significantly reduce axillary sweating in bromidrosis patients (Shelley 1951, 1954; Juhin and Hansson 1968). More effective odor prevention agents include antihydrotic (astringent) agents, which depress both apocrine and eccrine sweat by a constricting or binding effect which blocks sweat gland secretions.

Modern development of topical antiperspirants for axillary application can be dated back to Stillians' finding in 1916 that 25% aluminum chloride hexahydrate in distilled water, applied initially to the armpit every 2 or 3 days, will restrict excessive sweating (Stillians 1916). Stillians demonstrated that astringent lotions which contain zinc sulfate or alum (ammonium or potassium aluminum sulfate) usually relieved the condition after three applications, with subsequent use reduced to once a week or whenever needed. Aluminum chlorohydrate hexahydrate and other antiperspirant astringents such as formaldehyde or glutaraldehyde have allegedly been effective for relief of axillary hyperhidrosis, but have proved to be less effective in application to the palms or soles and may result in unsightly skin stains (Shelley et al. 1953; Strauss and Kligman 1956).

Other antihydrotic agents in use include aluminum salts (aluminum chloride, aluminum sulfate) or tannic acid. Alcoholic aluminum chloride hexahydrate mixture, applied on a daily basis, also seems to be effective (Jensen and

Karlsmark 1980). However, Knudsen and Neier (1963) reported that the anticholinergic drugs have been ineffective.

These astringent agents act to suppress apocrine and eccrine sweat. Their application is helpful to some extent in mild cases. In more extreme cases, however, the applied astringent is flushed away by copious sweat. Frequent application and strong concentrations of active ingredients may irritate the skin (Fig. 13.1). Aluminum salts, moreover, have a deteriorative effect on cotton clothing.

Sterilizer agents depress bacterial growth on the axillary skin surface, but cannot inhibit sweat gland secretions. Given that limitation, good results have been reported with methenamine (hexamine) (Cullen 1975). Adsorbent deodorants include chlorophyllins, ion exchange resins, and so forth. Many such adsorbent deodorants are sold in Japan, but are not truly efficacious.

13.1.2   Systemic Medication Therapy

In systemic medications, the use of anticholinergic drugs can be effective to some extent, but in severe cases the required doses result in negative side effects such as mouth and eye dryness. Some tests have been made to prevent sweat secretion by use of tranquilizers such as diazepam, reserpine, and antiautonomic nerve drugs. These drugs do not inhibit sweat secretion completely and can act to disturb the patient's motor balance, making it contraindicated to recommend them. Both tranquilizers and systemic anticholinergics, due to their side effects, show results even more distressful to the patient than the condition of excessive sweating.

13.1.3   Local Injection Therapy

Some physicians in Japan use local injection therapy to destroy the sweat glands. Typical agents include formalin (Imazu and Yokota 1931) and chlorophyllins. Kawabata (1930) and Yoshida (1987) developed a new type of injection therapy which may have utilized a derivative of a potassium cyanide agent. As evidence, hematuria occurred 1 or 2 days after the treatment, indicating that it caused toxification, and it was discontinued. Histological examination shows that the sweat glands are incompletely destroyed. Local injection therapy has therefore not proven useful. Yoshida (1987) presented a paper on this method but would not reveal the ingredients of the agent which he had developed for use in local injection. The follow-up results, however, proved that it is not efficacious.

Table 13.1. Frequency of application of various bromidrosis treatments

Total no. of patients treated	1972 635		1986–1987 1500	
Total no. of patients who had received prior treatment	124	(%)	234	(%)
No. of patients treated by:				
physical therapy	86	(69.4)	120	(51.3)
removal method	19	(15.3)	40	(17.1)
physical method and removal method	7	(5.6)	9	(3.8)
curettage method	7	(5.6)	14	(5.9)
clearance method	3	(2.4)	23	(9.8)
other	2	(1.6)	9	(3.8)
iontophoresis	0		0	
radiation therapy	0		0	
shaving method	0		19	(8.1)

(Reproduced from Inaba et al. 1988b)

13.2   Prior Treatments for Bromidrosis

Among 635 patients (Inaba 1986), 124 had already been treated by other procedures such as physical therapy (86 cases, 69.4%), an axillary skin removal method (19 cases, 15.3%), a removal method after physical therapy (7 cases, 5.6%), and a curettage method (7 cases, 5.6%), (Table 13.1). Iontophoresis and radiation therapy had not been used on any of these patients.

In a more recent study, 234 among 1,500 patients had received previous treatment in the form of electrocoagulation treatment in 120 cases (51.3%), removal treatment in 40 cases (17.1%), curettage in 14 cases (5.9%), and clearance operation in 23 cases (9.8%). The most interesting finding showed 9 cases of two other procedures (5 cases, suction method and 4 cases, injection method). This later report of 1988 (Inaba et al. 1988b), compared with the former report, shows that the rate of physical therapy treatment had declined (Table 13.1). This finding indicates that the patients perceived physical therapy as ineffective and tended to prefer the curettage and clearance methods.

Reoperation after subcutaneous tissue shaving occurred in 19 cases. In 3 cases, the patients had been treated in other clinics; in only 16 cases among 1,500, they had requested our procedure. In 3 cases, males wanted regrowth of axillary hair. For the sake of hair regeneration, shaving was incomplete. In the other 3 cases, reoperation was per-

Table 13.2. Relationship between the frequency of physical therapy and axillary hair regeneration

Average no. of treatments	No. of Cases	Results
5	10	Negligible regrowth (in 4 cases out of 10)
4	5	Almost no regrowth
3	34	Slightly less regrowth
2	25	Very slightly less regrowth
1	12	Normal regrowth
Total	86	

formed on only one axilla. These findings showed that our shaving method was very efficacious.

There were 86 patients who had received physical therapy. Among them, in 12 cases (14%), they had requested termination of treatment after one session because it was seen to be ineffective. After two sessions, another 25 patients (29%) terminated the therapy; after three sessions, 34 patients (40%); after four sessions, 5 more cases (5.8%); and after more than five times, another 10 cases (11.6%). Almost all of the patients had recognized drawbacks to the procedure and chose to terminate it (Table 13.2).

13.3 Physical Therapy

There are three types of physical therapies: (1) electrolysis and electrocoagulation and others, (2) iontophoresis, and (3) radiation. One can understand a choice of physical therapy without a full understanding of the options since pharmaceutical therapy is not too effective if the patient hopes to gain a good result without surgical procedures.

The odor of bromidrosis is in direct ratio to the numbers of apocrine glands appended to the hair follicles. If complete epilation is performed, this odor will be diminished by destroying the hair follicles, and thus the related apocrine glands. However, the condition of hyperhidrosis remains, because the ducts of the eccrine glands, which are not destroyed, open directly and independently on the skin surface. Patients remain concerned about sweat-soaked clothing in the axillary region. However, electrolysis and electrocoagulation can be useful for patients who have only slight axillary odor and no history of hyperhidrosis.

Use of an electrical current to remove hair was first put forward by Michel in 1875 with galvanic current to be used for electrochemical eradication of the hair follicle's germinative cells. Four years later, in 1879, Hardaway also applied this method to hair removal. A fine needle conducted galvanic (direct) current to chemically dissolve the hair root.

The next step in permanent hair destruction was taken in developing the method of thermolysis. This method provided for high-frequency electrocoagulation (diathermy) of the germinative hair cells and was successively recommended by Rostenberg (1925), Lerner (1942), and Erdos-Brown (1978). Since then, thermolysis has become quite sophisticated, and direct current methods have given way to use of alternating current. Hinkel and Lind (1981) pioneered the "blend" or "dual action" method in which galvanic and low-intensity, high-frequency currents were simultaneously used.

13.3.1   Electrolysis

In the galvanic (direct) current method a low-voltage, low-amperage current (30 V, 0.5–1.0 mA) is used to chemically ionize the tissues to which it is directed. The negative electrode is the electrologist's needle. The positive electrode is a dampened pad attached to the patient's body. Hydrogen gas and lye are generated within the hair follicle and the lye reacts with and decomposes the follicular tissue.

The galvanic current method is equally effective, apparently safer and less painful, but at the same time, much slower. The needle is inserted into a hair follicle as deep as 3–5 mm parallel to the hair shaft. In proper use, the needle should slide in easily. If pain or bleeding occur, the follicular wall has been pierced, indicating that the angle of insertion is wrong. Once the needle is inserted, the circuit is closed with a foot switch and current adjusted with a rheostat for a 0.2–2.0 mA of current flow. Hydrogen gas forms in tiny white bubbles around the base of the hair, which is then readily extracted from the follicle with forceps. Contiguous hair is not removed to prevent excessive inflammation (Wagner et al. 1985).

13.3.2   Electrocoagulation

High-frequency alternating current (thermolysis) has been the most popular method of permanent hair removal over the past 40 years. A mild epilating current can be produced by adjusting conventional electrosurgical machines of either the spark gap or vacuum space tube types. The needle

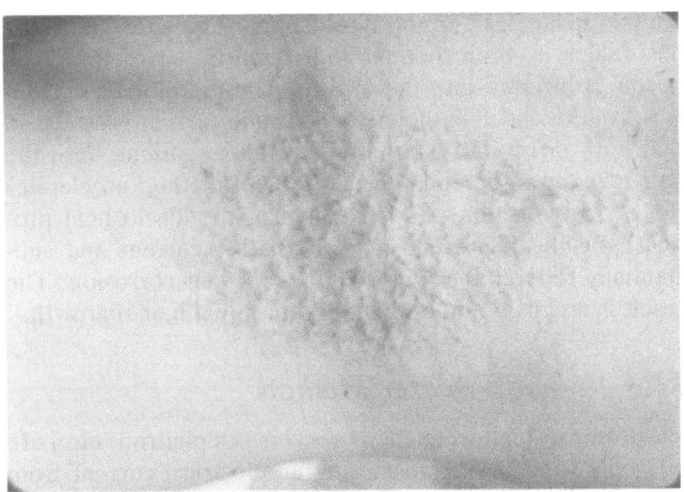

Fig. 13.2. Scarring caused by electrocoagulation. Electrocoagulation causes scarring more frequently than electrolysis if caution is not exercised

functions as the electrode. The water within the adjacent follicular tissue is highly conductive, and the action of the oscillating current heats the water, damaging the follicular tissue.

Thermolysis can be uniterminal or biterminal. The uniterminal technique (electrofulguration) causes tissue damage due to an electric energy transmitted by an electric arc or spark, using a dampened wave produced by a spark-gap circuit. Without using a ground plate, the energy goes through the patient. Two different methods are used in uniterminal thermolysis: manual and flash. The speed of the flash method, combined with relative lack of pain and good cosmetic result, make it the much preferred technique for epilation. Initial current duration and intensity settings are selected with due attention to the hair type and the body site to be treated. Once the settings are adjusted, epilation can be performed speedily (Wagner et al. 1985).

If the biterminal technique (electrocoagulation) is used, the tissue damage is largely due to heat (Fig. 13.2). The lesion is touched by the electrode. A ground plate is necessary. The electric energy is transmitted through the patient back to the machine (Swerdlow et al. 1974).

13.3.3 Blend Technique

The blend technique combines electrolysis and thermolysis in a single piece of equipment. This method has the apparent advantages of both techniques in subjecting the hair follicle to both thermal and chemical destruction. The needle is inserted into the follicle. The thermolytic current is activated, and 1–2 s later the galvanic current is added so that both direct and alternating currents continue until the hair is ready to be epilated. This blend method accelerates the corrosive action of sodium hydroxide due to heat produced by the high-frequency alternating current and substantially reduces the time required for hair corrosion. The result is less pain and only limited terminal hair regrowth.

13.3.4   The Tweezer Method

Electronic or high-frequency tweezers (Depilatron, etc.) are allegedly capable of transmitting an electrical current from the hair shaft into the hair follicle. However, there is no evidence that these electronic devices are beneficial, and they are not used at present in Japan.

We studied the effects this tweezer method would have on the treated hair follicles (unpublished work). Histological observation on the seventh postoperational day revealed atrophy of the sebaceous gland which had been caused by the transitory contraction of the vascular system after it received high-frequency waves. With the restoration of the sebaceous gland, regeneration of hair was observed at the upper isthmal portion of the hair follicle, a fact which confirmed that it never led to permanent epilation.

13.3.5   Light Flash Method

When the light irradiated from the Bio-Flash (Light Flash LF 100, YA-MAN Ltd., Tokyo) (Fig. 13.3) flashes in the hair follicle, the skin temperature rises by about 2°C. This results in the phenomenon of protein disintegration, making keratin disperse. Taking another example, when a proper temperature is applied to a gamic egg, the chicken hatches, but if the temperature applied is 1°–2°C above the normal degree, the protein disintegrates and the fetus dies. The Bio-Flash has been developed by applying a principle quite similar to this.

This newly-developed epilator has the following characteristics. It completely cuts off the ultraviolet rays that may bring harmful effects to the skin. Pain received by the skin is reduced to one-third of the conventional blend technique and no problems related to the treated areas are anticipated. The irradiated area is treated with a tweezer in a very short time, vastly minimizing the time required in the con-

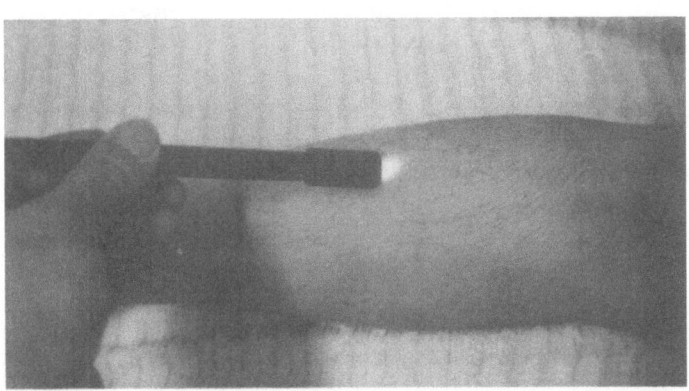

Fig. 13.3. The Light Flash Method. The instrument (Light Flash LF 100) is placed over the area to be treated. Light is irradiated from the tip and the skin temperature rises by about 2°C when it flashes in the hair follicle, resulting in protein disintegration

ventional method. The equipment only requires a very simple operation method. The producer claims that by applying an exclusively developed hair generation inhibiting agent, permanent epilation is practicable. However, it was difficult to obtain such permanent epilation. Hasegawa and Inaba (1989) reported that the follow-up results in clinical and histological research were not efficacious. This histological finding indicated only the atrophy of the sebaceous gland on the following day. The observation 3 weeks after the treatment revealed that hairs had not been epilated and the sebaceous glands had been restored, which was indicative of the inefficacy of this treatment. This method supports the authors' statement that only the destruction of the upper isthmal portion of the hair follicle leads to permanent epilation.

13.3.6   Review of Physical Therapies

We have conducted follow-up studies on some aspects of electrolysis or coagulation treatments. The major purpose of this was to determine the relationship between the frequency of these procedures and hair regeneration. A review of 86 patients who received the permanent epilation procedure in other clinics revealed that almost no regrowth was observed in 4 patients (5%), slight hair regrowth in 14 patients (16%), normal regrowth in 34 (39%), rather excessive regrowth in 28 (33%), and clearly excessive regrowth in 6 patients (7%) (Table 13.2). These findings

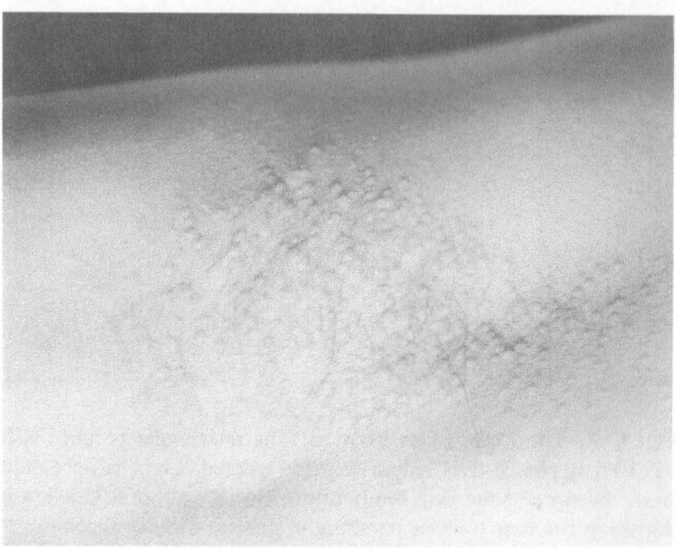

Fig. 13.4. Lichen planus papules observed in a patient with a keloid constitution

supported Savill and Warren's report "The limitations of electrolysis in skilled hands are those of cost and time; even the best operators can only deal with 25–100 hairs per sitting and hair regrows in up to 40% of the follicles treated" (Savill and Warren 1962). The conclusion must be drawn that physical therapy is inadequate for complete epilation of axillary hair. The greater the number of treatments, of course, the greater the epilatory effect. There is a risk of scar formation which is more prevalent in electrocoagulation therapy than in electrolysis. Other unpleasant complications include keloids, which are formed as lichen planus papules (Fig. 13.4).

We researched the degree of the odor after physical therapy. Only one bromidrosis patient of all cases surveyed in the patients' complaints had the odor completely diminished after physical therapy. On the other hand, in objective symptoms, in 17 cases (19%) the odor had completely diminished (Fig. 13.5). Excessive odor remained in 54 patients examined after these procedures. However, excessive hyperhidrosis remained in 78 patients (90%). Only one patient showed complete relief of hyperhidrosis (Fig. 13.6).

Physical therapy destroys the hair follicles in order to diminish the numbers of appended apocrine glands. According to our survey, however, hair regrowth and bromidrosis odor remain, because the upper isthmal portion of the sebaceous and the apocrine glands is not fully destroyed.

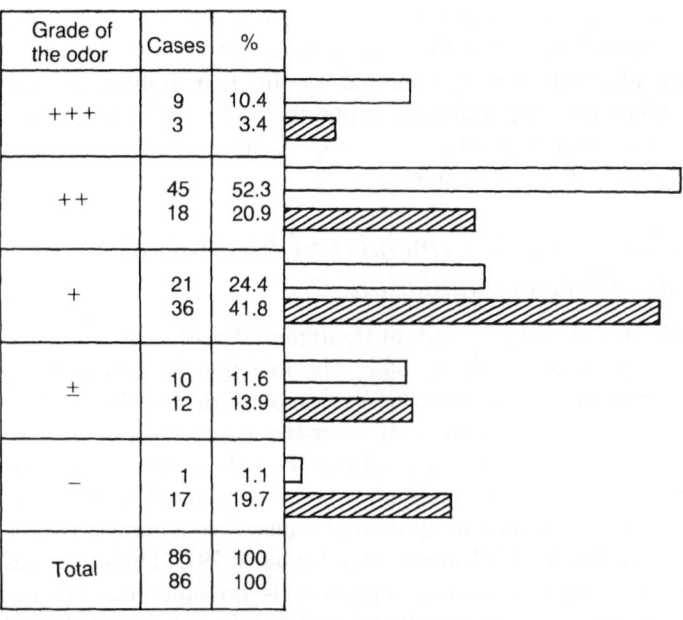

Grade of the odor	Cases	%
+++	9 3	10.4 3.4
++	45 18	52.3 20.9
+	21 36	24.4 41.8
±	10 12	11.6 13.9
−	1 17	1.1 19.7
Total	86 86	100 100

☐ According to the patient's complaints

▨ According to the physician (objective symptoms)

Degree of odor after physical therapy is very strong (+++); strong (++); moderate (+); faint (±); no odor (−)

Fig. 13.5. Degree of odor after physical therapy. Degree of odor after physical therapy is very strong (+++); strong (++); moderate (+); faint (±); no odor (−); (objective symptoms) (Reproduced with permission from Inaba 1986)

Hyper-hidrosis	Cases	%
+++	51	59.3
++	27	31.3
+	7	8.1
±	1	1.1
Total	86	100

Degree of hyperhidrosis after physical therapy is very strong (+++); strong (++); moderate (+); faint (±); no odor (−)

Fig. 13.6. Degree of hyperhidrosis after physical therapy. Degree of hyperhidrosis after physical therapy is very strong (+++); strong (++); moderate (+); faint (±); no odor (−). (Reproduced with permission from Inaba 1986)

The ducts of enlarged eccrine glands open directly and independently onto the skin surface and are not destroyed by physical therapy confined to the hair follicle, so the finding that hyperhidrosis is unaffected by these treatments is not surprising. Physical therapy is not useful in eliminating the odor of bromidrosis.

13.3.7 Basic Principles of Electrolysis and Electrocoagulation

Physical therapy is used for treatment of axillary bromidrosis by epilation of axillary hair. The fact that hairs frequently regrow in spite of supposedly complete destruction of hair roots has been attributed in many cases to a failure to destroy the hair, the dermal papilla, or too short hair roots in the telogen stage. Studies were conducted to examine supposedly complete epilation by electrocoagulation (Inaba et al. 1979b; McKinstry and Inaba 1979). These studies showed that permanent epilation is possible only by destruction of the upper isthmal portion of the hair follicle and sebaceous gland.

In order to confirm this finding, we investigated by destruction of those portions only, using oblique insertion of the needle at a 5mm distance from the hair canal (Fig. 13.7). If the upper isthmal portion remains intact after the lower portion of the follicle is destroyed, histological examination of hair regrowth after epilation by electrocoagulation reveals that new hair roots begin to form in the isthmal region of the follicle. On the other hand, if this isthmal portion is destroyed by electrocoagulation, the lower portion of the hair follicle rapidly changes into a premature telogen stage, finally to become terminal hair, in spite of the fact that the dermal papilla and hair bulb receive their blood supply from below. It has been thought that blood is supplied to hair follicles by the cutaneous blood plexus diverging from the dermis layer (transverse vessel branches) or by the vascular system in a form similar to a basket net which directly surrounds the hair follicles from the musculocutaneous arteries (Durward and Rudall 1958; Montagna and Ellis 1957) (Chap. 16).

The shock of destruction in the isthmal region may lead to a temporary imbalance in blood supply to the lower portions of the follicle following electrocoagulation. However, the long-range effect of complete disappearance of the hair root is unexpected, and this suggests that the dermal papilla is affected by the isthmus region in some obscure way related to blood vessels, nerves, or the sebaceous

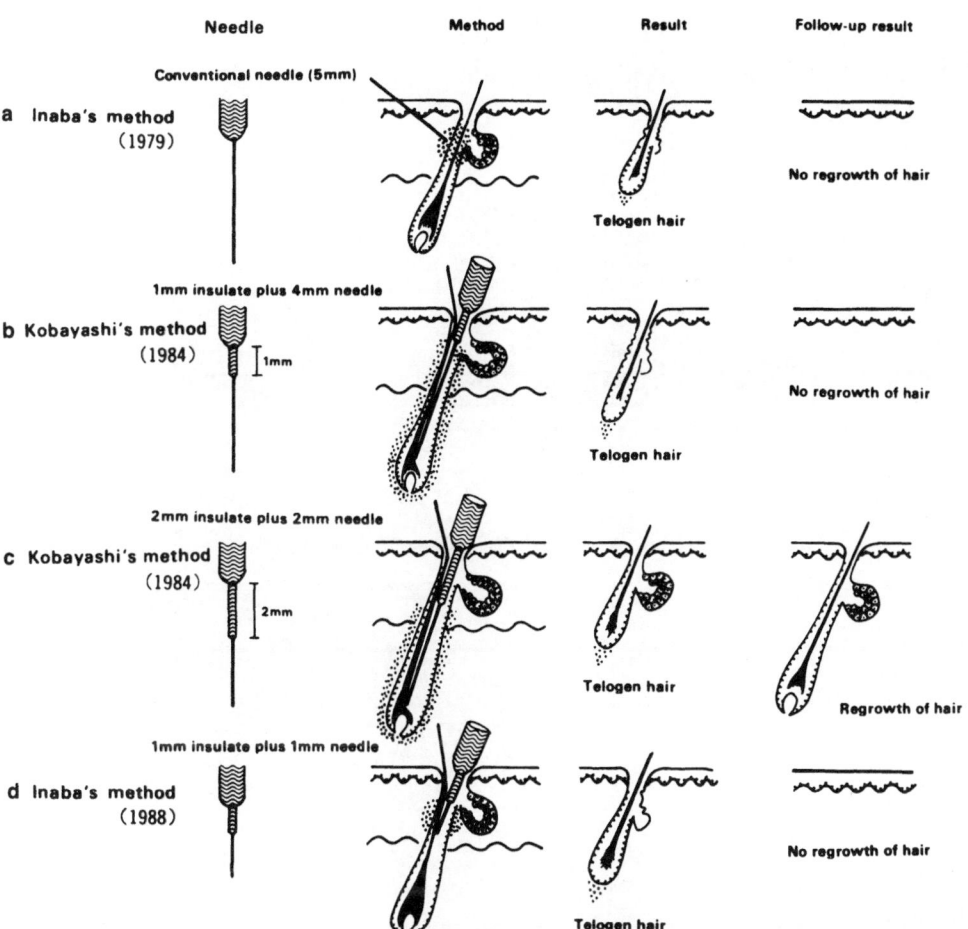

Fig. 13.7. Modifications in electrocoagulation needles used by Kobayashi and Inaba indicate the need to destroy the upper isthmal portion of the hair follicle to achieve permanent epilation. (Reproduced from Inaba et al. 1988b)

gland. This in turn suggests that the destruction of the follicular isthmus and the sebaceous gland by electrocoagulation is critical to permanent epilation of axillary hairs.

Hinkel and Lind (1981), in their exposition of electrolysis, note that the technique has relied on the assumption that destruction of the dermal papilla terminates hair growth, but they have not been able to point to specific research for confirmation. Their explanation of hair regrowth after electrolysis is curiously similar to ours in declaring that an entirely new follicle is regenerated after unsuccessful epilation and that its source is the outer root sheath of the

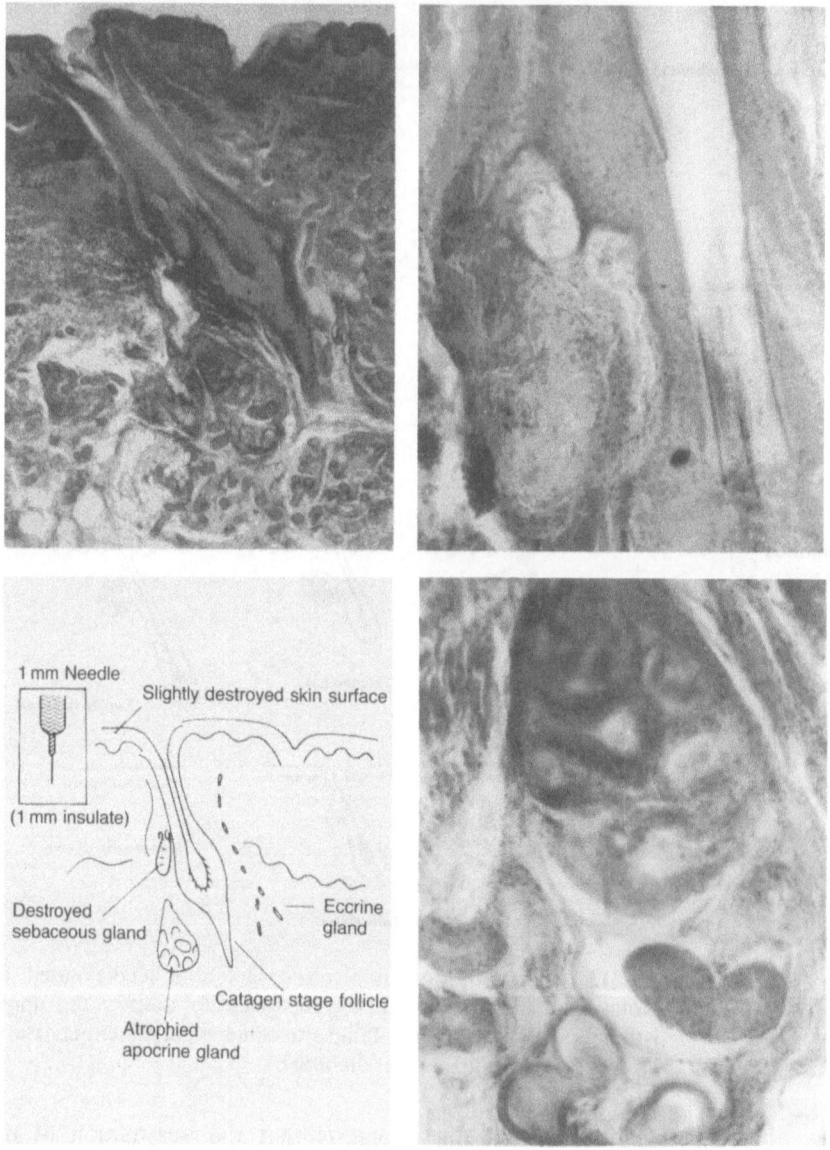

Fig. 13.8.a, b. A histological review of skin treated with a specially developed needle **a** Histological findings by electrocoagulation after 10 days using 1-mm insulate plus 1-mm needle. Sebaceous glands remain after using a special needle. In follow-up results, axillary hair regrows. **b** Histologic findings by electrocoagulation after 2 weeks using 1-mm insulate plus 1-mm needle. Sebaceous glands deteriorate after using the special needle. Axillary hair does not regrow

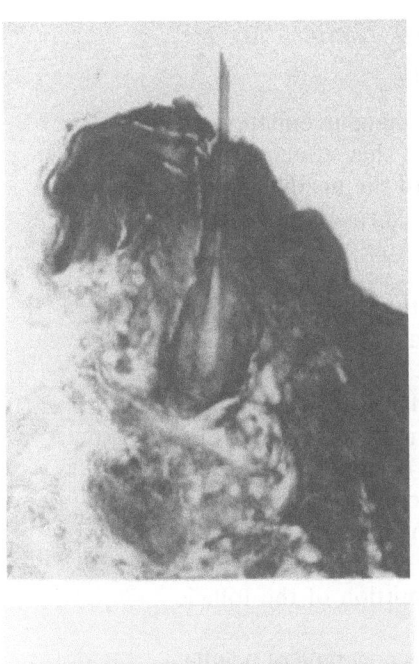
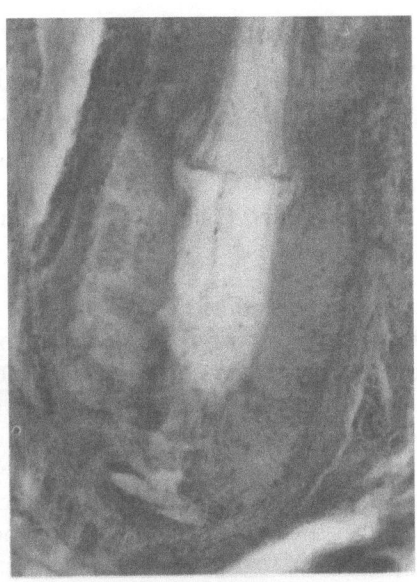
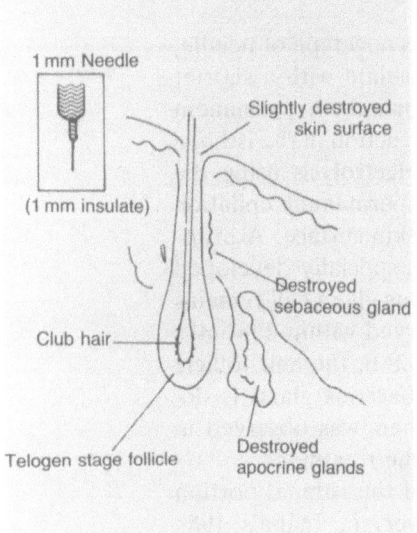
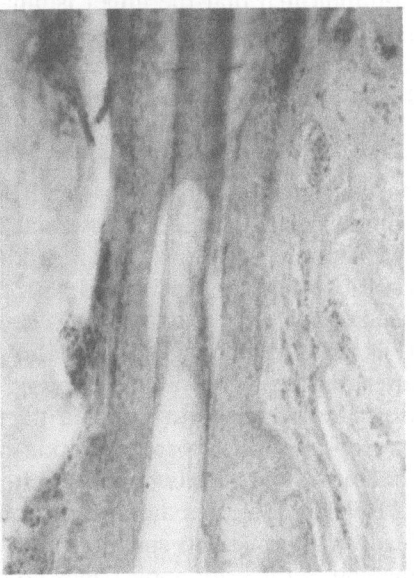

b

upper follicle (upper isthmal portion).

Barber and Jackson (1982) cite our hypothesis in their discussion of "Basic Principles of Electrolysis" in *Skin Surgery* as follows: "It appears that the previous theory requiring destruction of the papilla in order to destroy the hair may be inaccurate. These studies suggest that the proximal portion of the hair follicle including the isthmus must be destroyed to stop regeneration of the papilla and subsequent regrowth."

13.3.8 Supportive Findings for Basic Principles of Physical Therapy

Another report (Marton 1940) of permanent epilation by coated-needle electrocoagulation states that effective, permanent epilation requires positioning of the needle close to the skin surface. This, however, can cause small scars to form on the skin.

To perform non-scarring epilation, therefore, Kobayashi (1984, 1985) reported the use of a coated needle of a new construction (Figs. 13.7B, 13.9). If the coating is only 1 mm long (K-type needle), the needle will destroy the upper portion of the hair follicle (upper isthmal portion) for permanent epilation. If, however, the needle is coated to a length of 2 mm, the coating will protect the middle portion and the sebaceous gland, leaving both intact, with subsequent hair regrowth observed (Fig. 13.7C). This finding also indicated that the upper isthmal portion of the follicle is the essential center of hair regrowth.

In order to confirm the efficacy of this new type of needle, Inaba et al. (1988b) used the 1-mm insulate with a shorter needle which destroyed the isthmal portion for permanent epilation. As Fig. 13.7D shows, the destruction of the isthmal portion and the sebaceous gland by electrolysis using the 1-mm-insulated needle is essential to permanent epilation of axillary hair without damage to the skin surface. A histological review of skin treated with this specially developed needle shows that if part of the sebaceous gland cell remains (Fig. 13.8.a), hair regeneration is observed within 2 months after treatment. As observed in Fig. 13.8.b, the hair follicle turns to the telogen stage and the sebaceous gland is destroyed and lost. No regeneration of hair was observed in the follow-up result two months after the treatment.

This fact indicates that destruction of the isthmal portion leads to permanent epilation. However, in Inaba's 1988 method, it does not apply, since no immediate result of epilation is observed even after performing electrocoagulation.

13.3.9 Kobayashi's New Method for Treatment of Bromidrosis

In order to completely cure bromidrosis, Kobayashi (1988) developed a new treatment using two types of needles, one for electro-elimination of sweat glands and the other mainly for epilation (Fig. 13.9.a–d). The conventional method of permanent epilation aimed mainly to destroy the apocrine glands by using the K-type needle designed for permanent hair removal and for coagulation of sweat glands and sebaceous glands near the hair follicles. However, its shortcoming is that it fails to completely remove the eccrine glands, and the problem of hyperhidrosis remains.

Kobayashi then developed a new H-type needle. This needle has insulation on the upper half of the needle portion as well as on the 1-mm base, and is designed to electrically coagulate apocrine and eccrine sweat glands in the lower dermis and subdermis by inserting it parallel to the skin surface in the uppermost layer of the subdermis in an attempt to destroy the secretory coils. However, the essential point of this treatment is whether the central portion of the regeneration of the apocrine and eccrine glands (coiled ducts) present in the lower portion of the dermis is completely removed or not (Inaba et al. 1979b) (Figs. 15.15.a–e, 15.16.a,b). This indicates that the attempt to coagulate the secretory coils limited mainly to the subdermis layer leaves a problem of the regeneration of eccrine glands and cannot be fully recommended for the following reasons: (1) hyperhidrosis remains, (2) it causes burns resulting in necrosis, and (3) exudation becomes affluent, because it burns the subcutaneous tissue completely.

13.3.10 Problems of Physical Therapy

Many states in the United States require no licensing or certification for electrologists. The number of states which require a state examination and training is only 26 among the 50 states. In another 24 states, it is not under regulation (Wagner et al. 1985). The regulation code in California, the state at the forefront of enacting legal regulation of epilation by electrolysis, permits epilation by electrolysis to be performed only by qualified electrologists who must be over 17 years of age, who have graduated from the 12th grade of public school or who have equivalent scholastic certification, and who have completed a course of medical training (approved by the state beauty committee) for 500 h and passed the qualification exam. The problem of licensing requirements for electrologists is not confined to the United States. It is not fully approved as a medical practice in many countries, such as the EC member countries, Holland, Sweden, and so on. In the United Kingdom, although the services of electrologists are provided to patients through the National Health Service following the recommendation of a dermatologist, there is often hesitation to make such a

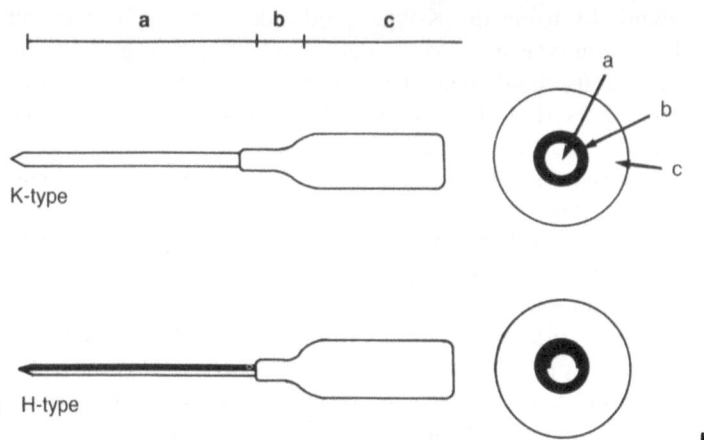

K-type

a H-type

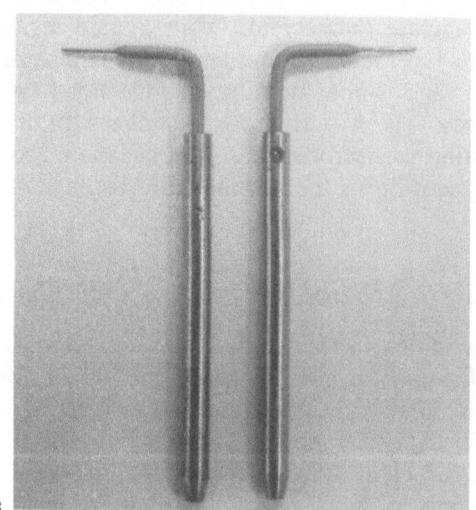

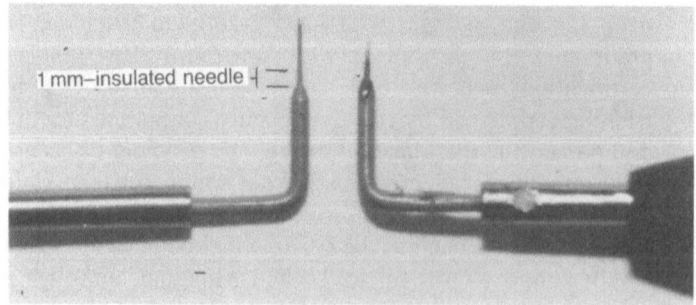

1 mm–insulated needle

referral on an ethical basis. This may be because in the United Kingdom there are no licensing requirements for those wishing to practice electrolysis (Ridley 1969).

In 1979, the Japanese government banned electrolysis at beauty salons. In spite of the ban, however, it is still performed privately at many beauty salons (Hayakawa 1988). Recently, there has been a move to amend the present law so that anyone with proper qualifications as an electrologist can perform it freely, as in many Western countries. That means separating medical epilation from aesthetic epilation and permitting electrolysis, which has relatively few side effects, to be performed at beauty salons and electrocoagulation only at medical facilities.

It is true that the results of permanent epilation performed at beauty salons seem fairly satisfactory, with no scars left on the treated areas. However, there are many patients who visit our clinic and confess that they have undergone this treatment for bromidrosis, firmly believing it would be a permanent cure. To their disappointment, hyperhidrosis remains and they begin to feel that they desperately need another treatment procedure.

13.3.11 Iontophoresis

As the process used to increase the penetration of an ionized substance into surface tissue by applying an electric current, iontophoresis has been utilized over the past 200 years to treat numerous disease conditions, including hyperhidrosis. In 1936, Ichihashi used various solutions of atropine histamine and formaldehyde to demonstrate that sweating of the palms could be reduced by iontophoresis. This method has won the most acceptance among dermatologists for treatment of hyperhidrosis, although the mechanism of

Fig. 13.9.a–d. Insertion methods of two types of insulated needles. The K-type needle has a 1-mm base insulation and the H-type needle has insulation on one side of the needle as well as on the 1-mm base **a** Enlarged view of the lateral sides of the K- and H-type needles. *a*, needle; *b*, 1 mm base; *c*, shaft **b** Enlarged view of the cross section of the K- and H-type needles. *Shadowed* parts indicate the insulated sections **c** H-type insulated needle has insulation on one side of the needle (*left*), while the other side (*right*) has no insulation **d** The K-type needle (*left*). In order to attempt to coagulate only the isthmal portion, the needle on the right has 1-mm insulation plus 1-mm needle

its action remains unclear. The most popular hypothesis is that mechanical blockage of the sweat ducts inhibits sweating at the stratum corneum level.

Selected ions are delivered into tissue by passing a direct current through a medicated solution. Sloan and Soltani (1986) state that the drug is applied under an electrode which has the same charge as the drug, and a return electrode opposite in charge to the drug is attached to a neutral site on the patient's body surface. A current below the level of the patient's pain threshold is selected and allowed to flow for an appropriate length of time. The electrical current gives a significant boost to the penetration of the drug into the surface tissue.

For specific treatment of hyperhidrosis, it was evident that sweat volume could be decreased by ion transfer of certain solutions applied to the skin. Tap water is the conducting medium most commonly in use because it is safe and effective (Grice et al. 1972; Abell and Morgan 1974; Shrivastava and Singh 1977). However, solutions of various compounds have been examined. Anticholinergic compounds such as poldine methyl sulfate (Hill 1976), glycopyrronium bromide, and atropine were found to have a longer lasting effect than water.

Bromidrosis odor is curbed for only 3 days by iontophoresis treatment. It is effective for treatment of plantar hyperhidrosis but not for treatment of axillary hyperhidrosis (Grice et al. 1972; Abell and Morgan 1974; Levit 1980).

13.3.12   Radiation Therapy

Bromidrosis has been treated by radiation, but a dosage which eliminates sweating also produces radiodermatitis (Borak et al. 1949; Cipollars and Crossland 1967). Radiation treatment can also cause ulcers. Long-term studies following radiation therapy for bromidrosis have revealed cases of cancer as well. At present, radiation therapy is contraindicated. In fact, the authors have not seen a patient who has undergone radiation therapy. It is much too hazardous in the doses required to atrophy the sweat glands.

Chapter 14. Surgical Procedures

Several types of surgical procedures have been developed for permanent relief of bromidrosis. These include surgical removal of axillary skin, surgical removal of subcutaneous tissue by the curettage method, the clearance method and other methods including the Inaba method. The ideal surgical treatment provides for good appearance after operation with only a small scar, full removal of the sweat glands, easy movement for the patient after the operation, and a short treatment period during hospitalization.

14.1 Radical Removal of Axillary Skin

Removal of axillary skin, including all the glands and hair follicles, is generally believed to be the most efficacious treatment procedure. In that context, it should be noted that the length and width of the axillary hair-bearing region is influenced by hormonal activity. The average width of this area in Japanese females is 4 cm and in males 7 cm. The average length is 8 cm for females and 10 cm for males.

Complete removal of the axillary hair-bearing region is impossible in practice because it is too large. Suturing so broad an excision causes a dog-eared appearance (Fig. 14.1). There are no problems concerning the length, but the width gives rise to stitching difficulties. A suture is usually removed 7 days after an operation. In this case, however, suture removal may require a wait of 2 weeks, given the risk that the skin will be stretched and the cut could open again. Accordingly, healing time is that much longer.

With patients who have a keloid constitution, contractions may occur and cause complications of mobility. After the suture is first removed, the scar is a line, but after some

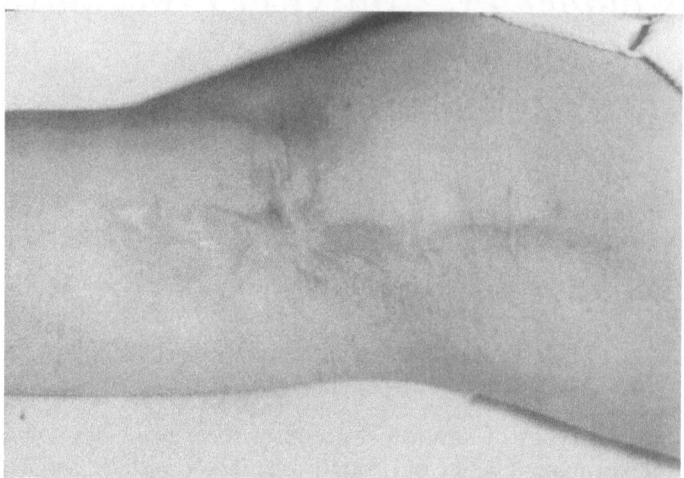

Fig. 14.1. Large scar left by complete excision method to eliminate perspiration

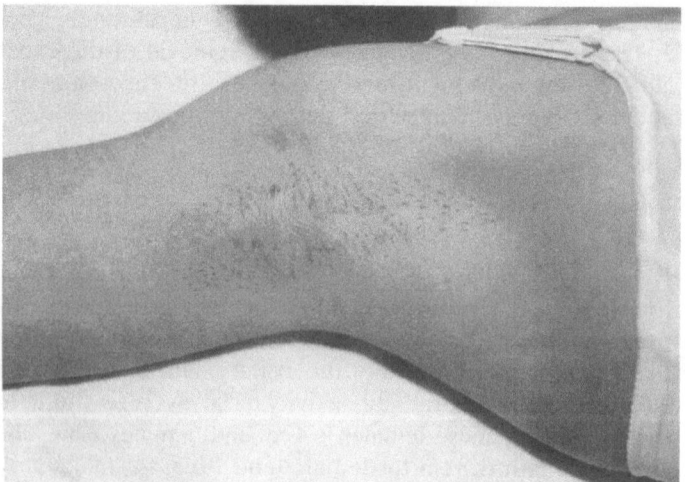

Fig. 14.2. Incomplete excision method. After the suture is removed, a linear scar is observed

time, because of the patient's arm movements, the scar continues to widen until it reaches the original width of the skin removed. To avoid this complication, the physician is restricted to making an incomplete incision.

Incomplete removal involves only a part of the hair-bearing axillary skin. The surgical scar is much the same as that of the former procedure, and portions of axillary hair remain (Fig. 14.2).

Hurley and Shelley (1963) initially performed a simple elliptical excision through the center of the major per-

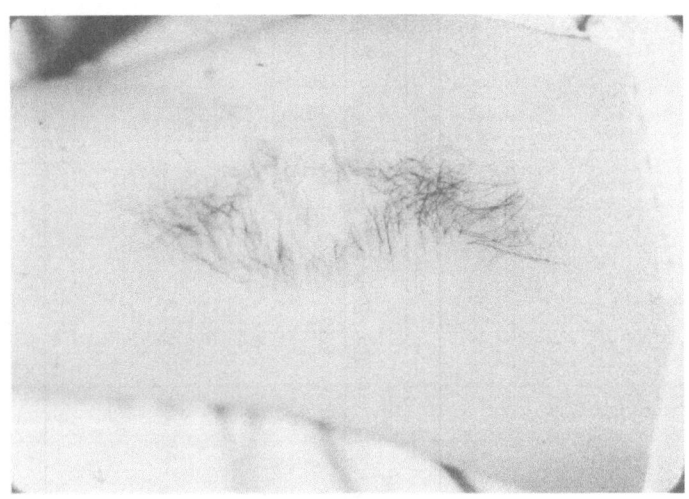

Fig. 14.3. Incomplete excision. Expanded scar following incomplete excision surgery. Axillary hair remains outside the scar area

spiration zone, usually equivalent to the axillary center. This method reduced sweating by 80% and achieved a good subjective result due to the fact that hyper-responsive glands tend to be clustered in a circumscribed control area. However, some patients were reportedly not pleased by the newly restricted arm movement (Ellis 1975).

Hurley and Shelley later (1966) improved this method to include elliptical excision plus edge undermining. Nevertheless, the resultant scar by this method is poor in appearance, similar to a bald scalp surrounded by a "tonsure" of remnant axillary hair (Fig. 14.3). Incomplete removal operations do not provide complete relief from bromidrosis and cannot be recommended.

Ozaki et al. (1953) reported that in 40 cases which involved total removal of axillary skin, the excellent result rate was 19%, fair 57%, and poor 24% (Table 14.1). Bretteville-Jansen (1973) reported on 16 cases of partial removal in which the S-shaped excision was adopted. Excellent follow-up results were observed in 25%, while poor results were seen in 37.5%.

14.2   Other Removal Methods

Z-plasty and W-plasty have been used to reduce the scar area (Figs. 14.4a–c).

This is a common technical procedure and one of the

Table 14.1. Excision methods: Follow-up results by updated treatment

Method	Developer	Reporter	Operation method	Cases	Treatment days	Complication cases (%)	Cases	Follow-up results (%)			
								Excellent	Very good	Good	Poor
Complete removal		Ozaki et al. 1953	Total excision	40	More than 2 weeks in 30%	14 (20.9)	40	19		57	24
		Kuroda 1957		52			11		100		9
		Ishida 1958		57					91		
		Murata 1960					49		75.5		24.5
Partial removal	Hurley and Shelley 1963	Bretteville-Jansen 1973	Limited excision	2	21	2 (100)	2			50	50
			S-shaped excision	16	20	4 (25)	16	25		37.5	37.5
Z-plasty	Greeley 1950	Bretteville-Jansen 1973	Total excision	21	15	3 (14.3)	21	76.1		19.2	4.7

(Reproduced with permission from Inaba 1986)

a Principle Z-plasty (T)

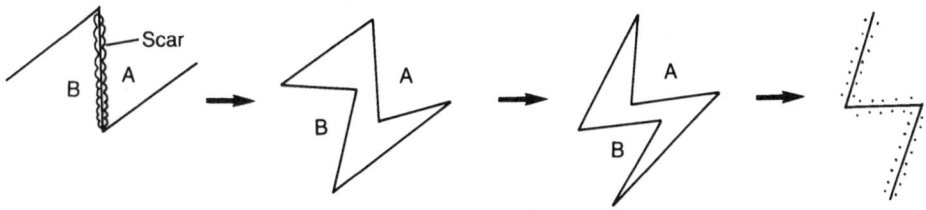

b Z-plasty method for the removal operation

c W-plasty (Double-plasty) method

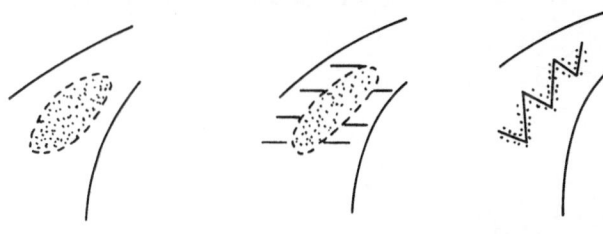

Fig. 14.4.a–c. The Z-plasty procedure for reducing the scar area. Z-plasty (a transposition flap) is designated as such because it is a technique by which two triangular flaps (*A* and *B*) outlined as a Z are interchanged one for the other. (Reproduced with permission from Inaba 1986)

most effective in plastic surgical repair. The principal uses for Z-plasty are: (1) to increase the length of skin in a predetermined direction, (2) to change the alignment of the scar to a more favorable direction which follows preexisting skin lines, and (3), after total resection, to add A and B triangular flaps (Fig. 14.4.a) which are then exchanged. The purpose of this technique is to make the scar less visible, e.g., the scar is concealed in an axillary skin wrinkle so that less expansion or retraction occurs. The drawbacks of this technique include the large incision, the large scar, and many remnant axillary hairs (Figs. 14.5, 14.6).

Bretteville-Jansen (1973) recommended total excision of the axillary perspiration area followed by Z-plasty closure

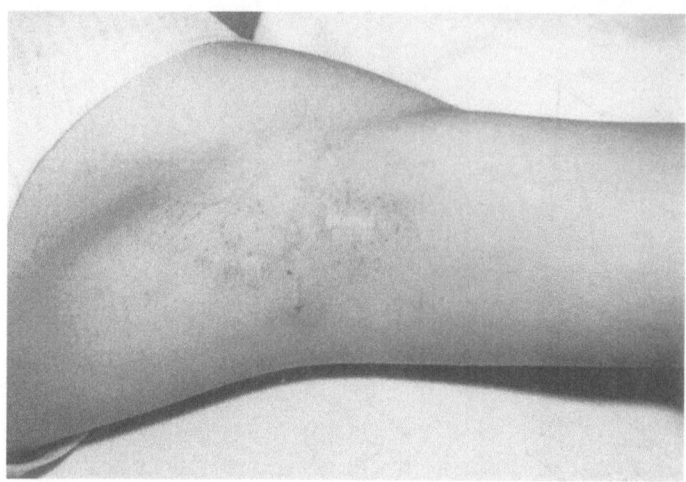

Fig. 14.5. Appearance of axilla after incomplete excision by Z-plasty method. The drawbacks of this technique include the large incision, the large scar, and many remnant axillary hairs

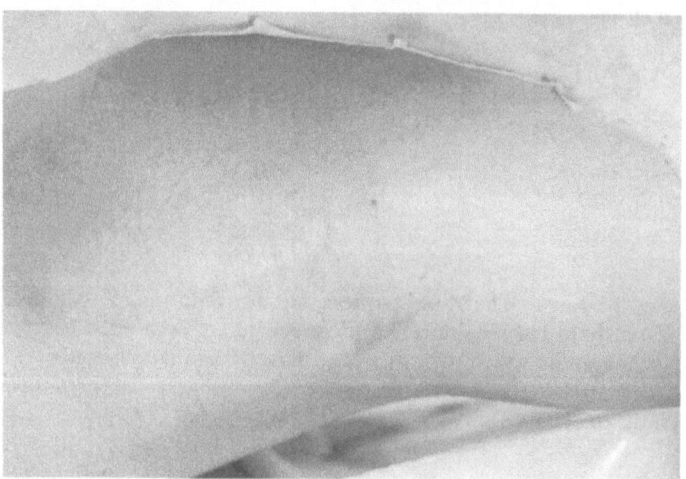

Fig. 14.6. Case two of the Z-plasty method

to ensure complete relief from hyperhidrosis. The twofold advantage of the Z-plasty closure is the elongation line and distribution of tension, partly across the suture line over a longer distance and partly as a shearing strain. However, this method is still not radical enough and leaves a visible scar (Table 14.1).

W-plasty (double Z-plasty method) is another method used to diminish the scar. The technique is to make more than one small Z incision (Fig. 14.7). In general, both

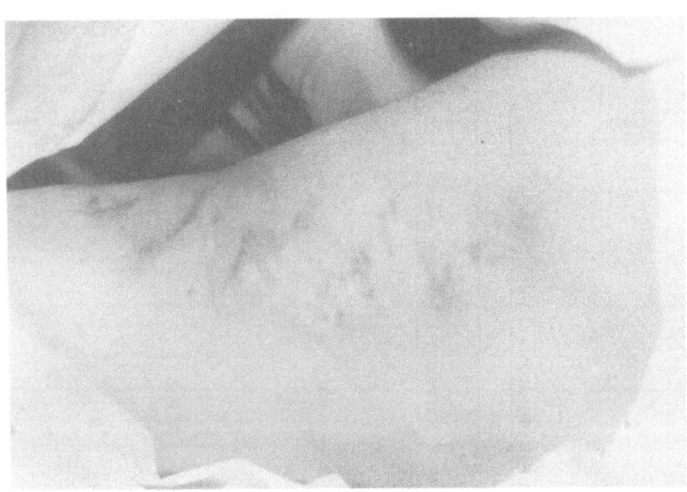

Fig. 14.7. Example of serial W-plasty method. This method results in large, ugly scars

Z-plasty and W-plasty cause large and ugly scars, because of the wide excisions required. Due to the disadvantages, these techniques are not used anymore.

14.3   Curettage Method

In 1948 Kano (1952) introduced a new curettage method. With a small incision of 2–3 cm and using a curette (axillary skin not removed), the tissue was curetted, with the hand used for counterpressure. However, complete removal of eccrine glands and hair follicles was impossible, and the result was not encouraging. Kano reported (Table 14.2) that out of 135 cases, 89% were good results, 11% were failures and 6% had complications.

Jemec (1975) developed a basic axillae-subcutaneous curettage method. The marked skin area is undermined with a pair of scissors through one or two short incisions. A sharp gynecological curette is employed to remove all subcutaneous fat in the area demarcated by the starch-iodine test. Curetting is continued until small petechiae appear on the skin surface, After approximately 10 min of scraping, the wound is drained, and the small incisions are sutured; the stitches are removed in about 10 days. In a report on 20 patients treated with this procedure and evaluated 6–9 months later, Jemec was informed that 12 (60%) were fully satisfied, 4 (20%) partly satisfied, 3 (15%) dis-

Table 14.2. Subcutaneous removal methods. (Reproduced from 'Treatments for hyperhidrosis and bromidrosis' Inaba, 1986)

Method	Developer	Reporter	Operation method	Cases	Treatment days	Complication cases (%)	Cases	Follow-up results (%)				
								Excellent	Very good	Good	No good	Not clear
Curettage	Kano 1948	Kano 1952		135	More than 2 weeks in 40%	8 (5.9)	135		88.9		11.1	
		Kuroda 1957		30		18 (60)	21		100			
		Nakahira 1957		48			48		66.5		33.5	
		Jemec 1975		20			20	60	20		15	
		Ellis 1975		12			12	17		8	75	5
Clearance	Tanioku 1952	Tanioku 1956 Fumiiri 1968 Harahap 1979							85–90		10–15	
	Skoog and Thyreson 1962	Skoog and Thyreson 1962	Z-incision	15		3 (20)	15	80		20		
		Bretteville-Jansen 1973	Off-set cruciate incision	7	30	6 (85.7)	7	29		14	57	
		Bretteville-Jansen 1973	S-shaped incision	15	25	9 (60)	15	20		33	47	
Sub-cutaneous shaver	Inaba and Takagi (1971)	Inaba 1976	External axillary incision	456	7	14 (3)	202	92.6		4.5	2.9	

(Reproduced with permission from Inaba 1986)

satisfied, and 15% uncertain (Table 14.2). One patient did not complete follow-up. None of the patients had complaints related to scars. A similar number of patients receiving excision surgery indicated fairly similar results, but four of them reported problems associated with the axillary scarring.

Encouraged by this report, Ellis (1975) decided to try out the procedure. His report indicates that 2 patients (17%) in 12 cases were thoroughly satisfied. One case (8%) showed a good result on one side but no effect on the other, and of the remaining 9 cases (75%), one reported being 50% improved, one patient completely failed to respond to the operation, and the remaining 7 relapsed completely from 1–5 months after curettage. Five of these had already been reoperated upon by the axillary skin excision technique and one was on the waiting list for surgery. Ellis (1977) later reported that this curettage method proved to be inadequate. As indicated in Table 13.1, of the patients who had received prior treatment, 6% had been treated by the curettage method. Most of these patients complain of regeneration of the axillary hair and no improvements in bromidrosis and hyperhidrosis. Therefore, this method is still inadequate from the standpoint of truly effective surgical treatment (Fig. 14.8).

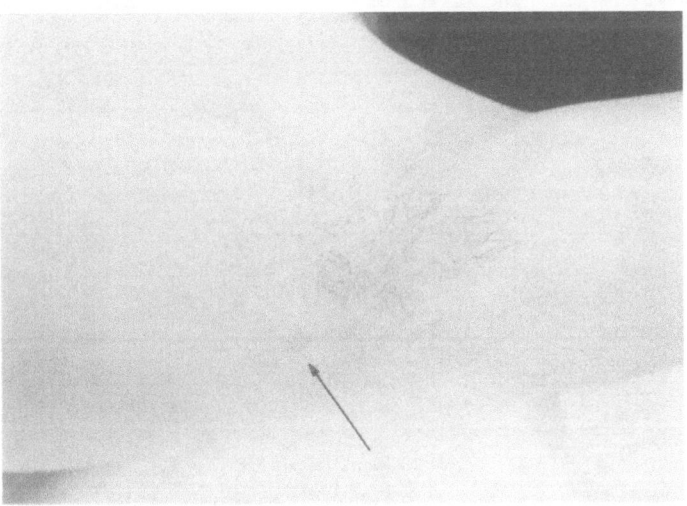

Fig. 14.8. Appearance of axilla after curettage method. The *arrow* indicates the incision site. Complete removal of eccrine glands and hair follicles is impossible. The result is not encouraging

14.4   Clearance Method

In this method (Figs. 14.9.a,b), the incision is made along
a wrinkle. The skin is turned over and pushed up by the
fingers. Sweat glands and hair follicles on the inner side are
removed with scissoring to leave the skin in thick-split
thickness.

One incision is not sufficient to remove all of the sweat
glands and hair follicles, especially on both of the lateral
sides (Figs. 14.10.a,b). A large incision is needed for complete removal. At present, aesthetic surgeons in Japan

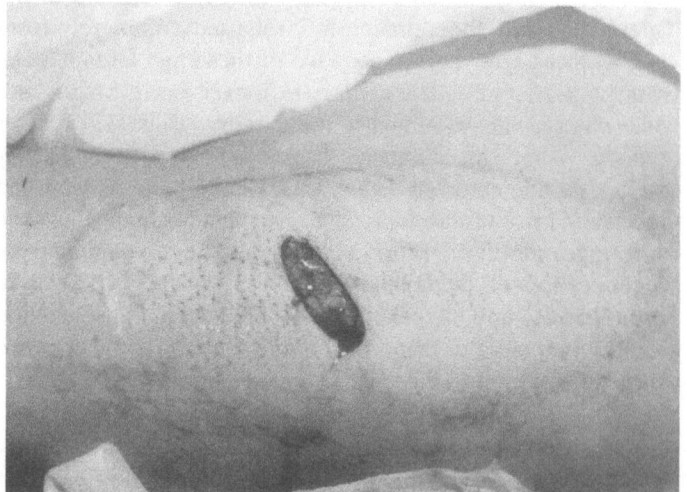

a

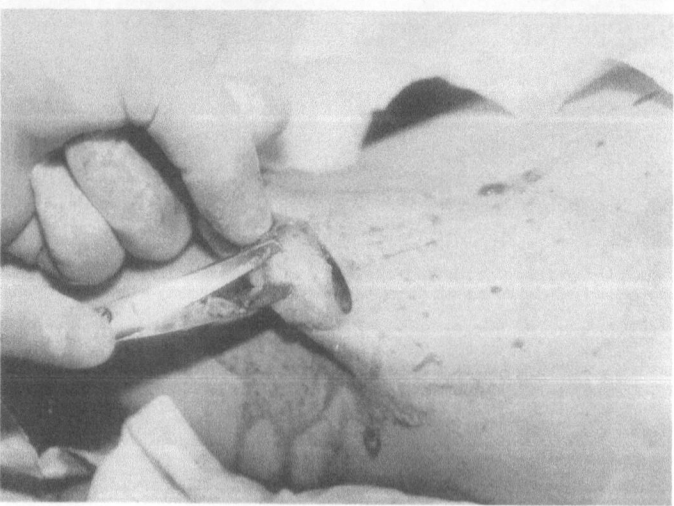

b

Fig. 14.9.a,b. The clearance method **a** Incision is made along the
line of wrinkling **b** Snipping is employed for removal of sweat
glands and hair follicles

make two incisions according to the length of the axillary region (Figs. 14.11.a,b). The originator of this method is not known, but the method as practiced in Japan has not been reported in the international medical literature.

Harahap (1979) reported a similar double clearance method, with two transverse parallel incisions made in the skin creases across the axillary fossa and down to the subcutaneous fat. The incisions are about 5–6 cm long and about 4–5 cm apart. The edges of the incisions are elevated to expose the tissue which bears the sweat glands and hair follicles. Subcutaneous tissue beyond the two incisions and under the bridge of skin between them is removed in depth,

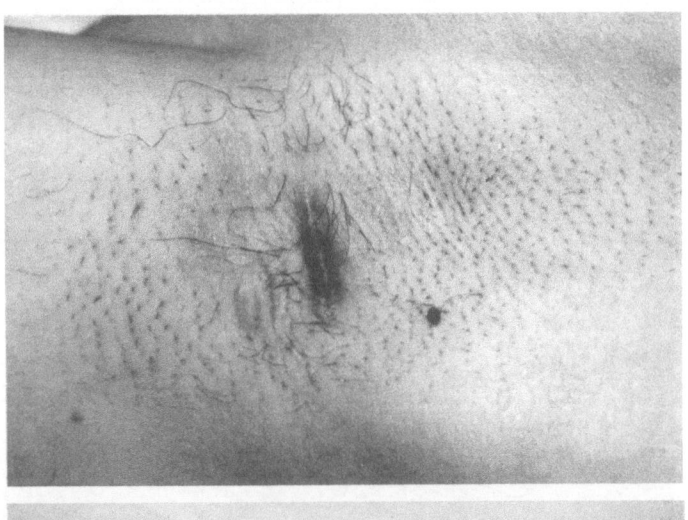

a

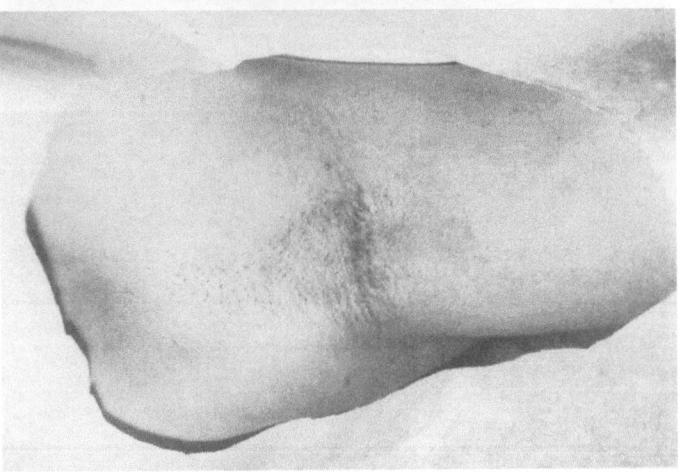

b

Fig. 14.10.a,b. The clearance method. A single incision results in incomplete removal of sweat glands and hair follicles. A large scar remains

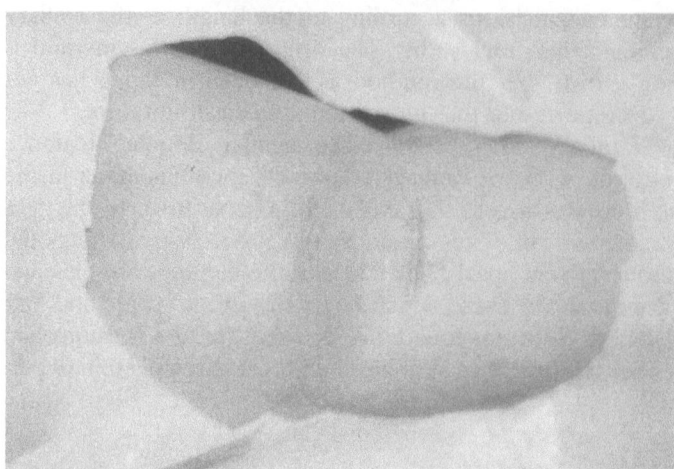

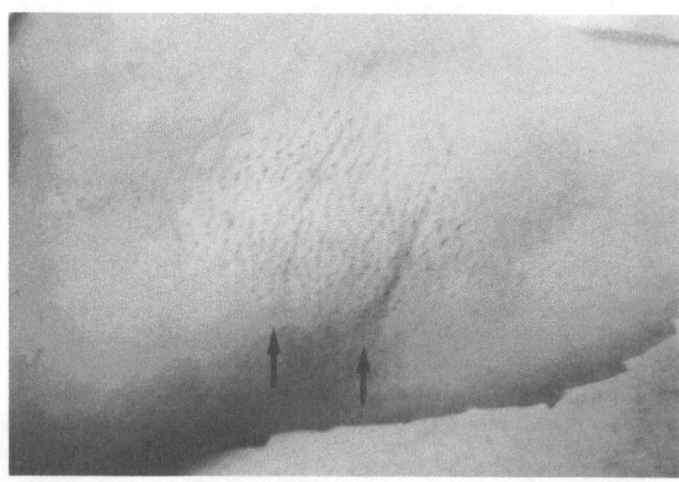

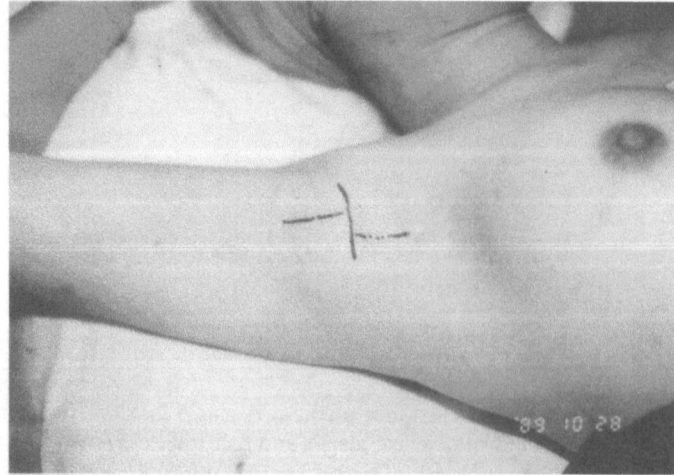

but in a way that leaves the skin in full thickness.

Skill is required to remove the sweat glands and hair follicles completely. If done ineptly, skin damage can occur, and if the patient has a keloid constitution, the scar is ugly. The operation time is 1–2 h and the cost of this procedure is high.

This method is sound for staunching bleeding. Direct hemostasis is done by using electrocoagulation. Bleeding may sometimes cause hematoma, but dressing is easier. In the immediate postoperative period anhidrosis and hypoesthesia may be observed. Afterwards (in about 3 months) sweating may resume to some degree.

The Skoog method (Skoog and Thyreson 1962) was another method of radical gland clearance with vascular flap (Fig. 14.11.c). It deals with the entire sweat gland-bearing area through an off-set cruciate incision (Table 14.2). The four flaps are lifted with the attached sweat glands, which are then snipped off before the flaps are sutured back in position. This technique preserves the subcutaneous network of blood vessels for good circulation within the flaps. Axillary sweating was abolished for a period of 2–3 months, but resumed in a normal pattern later on.

Based on removal of the apocrine glands (subcutaneous), this method leaves the dermis, epidermis, and subpapillary vessels intact, as well as the eccrine glands which are situated at the junction between the dermis and the subcutaneous fat layer or sometimes in the lower third of the dermis. On the other hand, with the central portion of the hair follicle found to be situated at the upper isthmal portion, there is axillary hair regrowth. Still, the technique did have value in maintaining good blood circulation to the skin (Yano et al. 1988).

In populations which have high numbers of apocrine glands, this technique can remove only the secretory coil of the apocrine glands in the subcutaneous layer, while eccrine glands and hair follicles remain, so that hyperhidrosis and hair growth persist. This technique has been demonstrated to be less than perfect.

Fig. 14.11.a,b. In order to remove the sweat glands in the bilateral region, physicians have used two incisions along wrinkle lines. A large scar remains (*arrows*) **c** The Skoog method of cruciate incision. The four flaps are lifted with the attached sweat glands, which are then snipped off. This method is not efficacious because the eccrine glands are not removed

14.5   Removal and Clearance Method

Following a partial elliptical excision, the edge of the skin
is turned over and the inner side cleaned with scissors.
Guerrero-Santos (1971) developed a technique of elliptical
excision and subcutaneous clearance with preservation
of vascularized flaps. Rigg (1977) reported a variation of
simple elliptical excision plus radical gland clearance with
conversion of edges from flap to skin graft. Many other
procedures have been attempted to diminish the remaining
scar by using the partial removal plus clearance method
in which edge undermining is done with subcutaneous
clearance. If the subcutaneous clearance is complete,
however, the suture is bound to reopen. Later, after the
suture is removed, the scar is widened. This method is
not recommended because of the large scar and lack of
efficiency.

14.6   Clearance Method and
Electrocoagulation of Sebaceous Glands

Nishimura et al. (1986) developed another new clearance
method which obtained about 50%–60% improvement in
bromidrosis related to sweating, odor, and regeneration of
axillary hair. The new treatment used the clearance method
plus electrocoagulation (bipolar method) of the sebaceous
gland and hair follicle. This method has been efficacious but
remains incomplete (70%–80%), because the eccrine
glands involved in hyperhidrosis remain intact. In fact, 3
patients out of 23 who had received this treatment method
during 1985 and 1986 came to our clinic to undergo a
reoperation.

The shortcoming of this method is that since it completely
burns the sebaceous glands, exudation becomes affluent.
The patient is required to be hospitalized for about 2 weeks
for draining the exudation by hemopack in order to avoid
cyst formation. For this reason application of this method is
not advisable.

14.7   Cervicothoracic Sympathectomy

Sympathectomy is not an easily recommended procedure
for volar hyperhidrosis of the palm and axillary hyperhidrosis. The first method described to accomplish this was
reported by Adson and Brown (1932). Using a posterior

approach, the second or third rib was exposed and 4–5 cm of the ribs excised to gain access to the sympathetic chain. Telford (1938) described the supraclavicular approach which has remained the most popular method.

The sympathectomy described by Atkins (1949) using a peraxillary transpleural route, seems in almost every way superior to the other sympathectomies. Among its disadvantages is a compensatory thermal hyperhidrosis, principally of the trunk, or a compensatory gustatory hyperhidrosis that may ensue (Shelley and Florence 1960) (Chap. 12.1).

14.8   Other Methods

In the grafting method, the skin of the axillary hair region is removed, and a skin graft is taken from some other part of the body. A large scar is left at the donor site, and the suture line at the recipient area and different skin pigmentation remain visible (Fig. 14.12.a,b).

Yoshida (1963) reported a radical surgical treatment for bromidrosis using a dermatome. He developed a treatment which collects the underarm skin of the patient up to the thickness of approximately 1/50 in. (0.5 mm) by using a Padgett's dermatome (1939) for preservation. The deep dermal layer and the remaining subcutaneous tissue layer are then removed from the area and the medium split-thickness skin graft is grafted over the area. Better results can be obtained by using this method, compared with various radical operations. However, its shortcomings are that pigmentation and contraction of the skin graft may occur.

Sho (1969) reported that he had obtained a satisfactory result by using a pedicle flap. This method is to form a U-shaped pedicle flap in the underarm skin, remove the sweat glands from the reverse side, and graft over the area where it was originally removed. Since this method assures hemostasis, the pressed tie-over can be done easily with successful results. However, it has the shortcoming that the scar will be enlarged.

Still other methods used to damage the sweat glands include: (1) cryotherapy (Ashby and Williams 1976), (2) a subcutaneous tissue shaver introduced by Landes and Kappesser (1979) which employs a housed razor blade to remove subcutaneous sweat glands (but not completely, and leaves a large scar), and (3) a new liposuction curettage approach using a special cannula (Shirakabe et al. 1986;

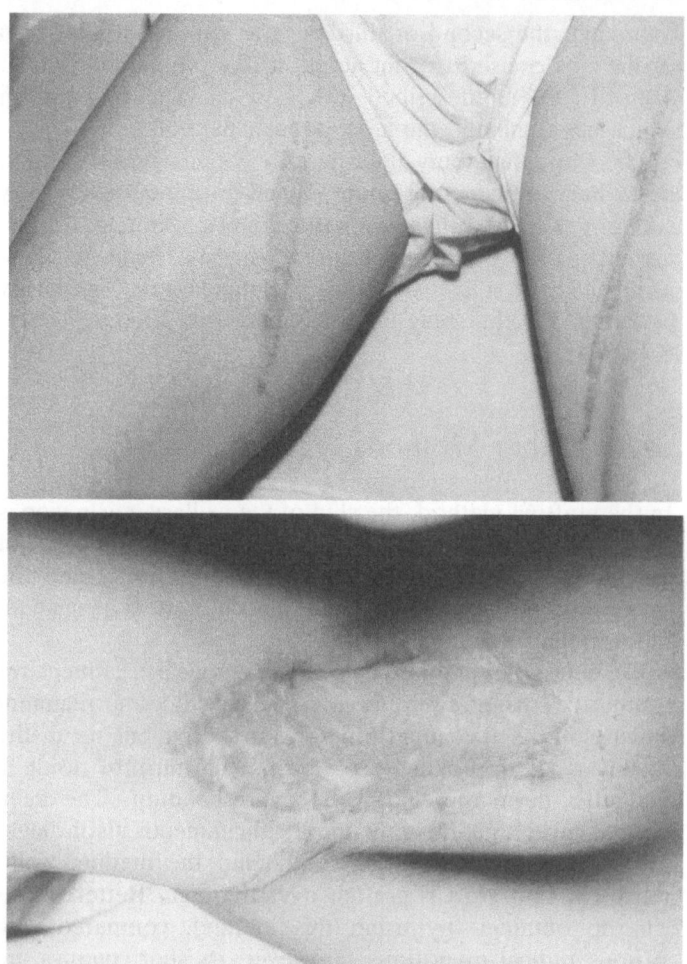

Fig. 14.12.a,b. The grafting method **a** Graft donor sites on inner thighs. A large, ugly scar remains **b** Graft implantation on axillary region. An ugly scar remains

Shenag and Psira 1987; Crisostomo 1989), which is not efficacious because the sweat glands regenerate (Fig. 14.13). Crisostomo (1989) reported that the 50 cases studied have all been successfully treated except for 2 patients in whom some recurrence has been reported. However, there are 5 cases of patients who underwent the liposuction curettage method without any effective results and visited our clinic for reoperation. They complained that they were not satisfied with the operation because hyperhidrosis persisted and axillary odor recurred with the passage of time. Therefore, this method is not considered recommendable. Many

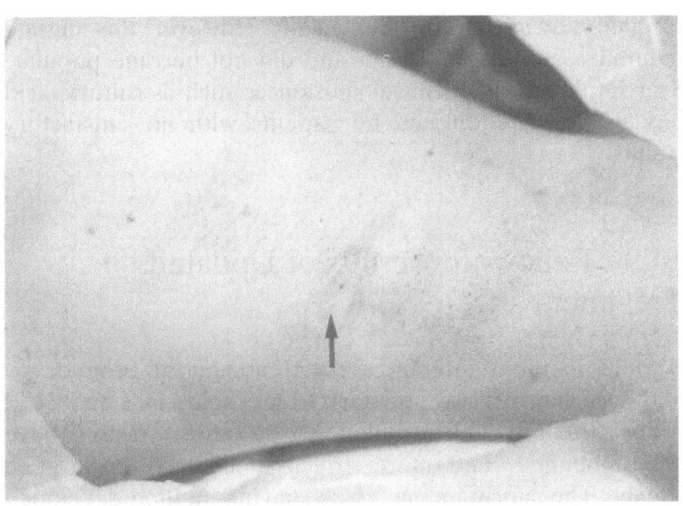

Fig. 14.13. Liposuction curettage method. This method is not efficacious. Regrowth of axillary hair and sweat glands is observed. The *arrow* indicates the incision site for inserting the cannula

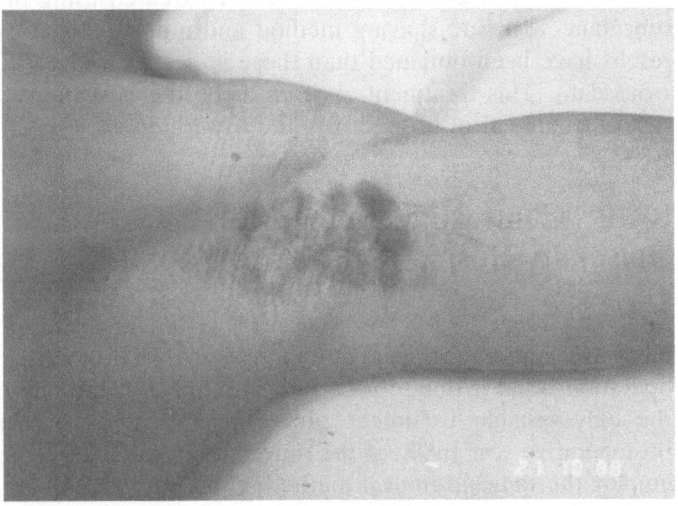

Fig. 14.14. An axilla bearing marks of moxa (burn) treatment. The treatment was not completed due to the pain caused to the patient

methods of radical surgery designed to deal with the localized problem of axillary hyperhidrosis carry significant risks and side effects which exceed the symptomatology (Hartfall and Jochimsen 1972). In some such cases the patients, in their desperate hope to diminish the axillary odor, attempted to destroy the sweat glands by applying scar formation by moxa burn (Fig. 14.14). However, this method naturally caused severe pain and did not become popular. Scar formation by chemical substances such as sulfuric acid has also been attempted by patients with no satisfactory results.

14.9   Follow-up Results of Updated Treatment

Various methods attempted for treatment of bromidrosis and hyperhidrosis are summarized in Tables 14.1 and 14.2. These attempts indicate that no successful results have been obtained, and more effective methods have been sought. The subcutaneous tissue shaving method developed by the author, which will be described in detail in Chap. 15, can be considered the best method so far, from both the aesthetic and efficacious standpoints. Table 14.2 shows the results of a questionnaire conducted on 456 cases. Up to the present, over 20,000 cases have been treated using the subcutaneous tissue shaving method and more satisfactory results have been obtained than those indicated in the previous data. This treatment appears to be the best one for radical treatment of bromidrosis and hyperhidrosis.

14.10   Contrasting Methods of University Hospitals and Private Clinics

In Japan, general surgeons, especially in university hospitals, have generally not paid attention to scars in bromidrosis operations, and assume that a radical removal technique is the only reliable treatment procedure regardless of the postoperative scar (68% of the university hospital surgeons employ the radical removal method, Z-plasty and W-plasty methods) (Fig. 14.15). They believe the clearance method may result in the regrowth of sweat glands from the lower end of the shaved sweat duct, and thus prefer radical removal. However, the prospect of an overlarge scar restricts them to partial removal, which does not satisfy the patients (Inaba 1986).

Patients tend to assume that the methods used in university hospitals are necessarily the best, but upon learning that scars are larger and hospitalization is longer, they choose instead to be treated in private aesthetic surgery clinics. Reports from one group of university hospitals over

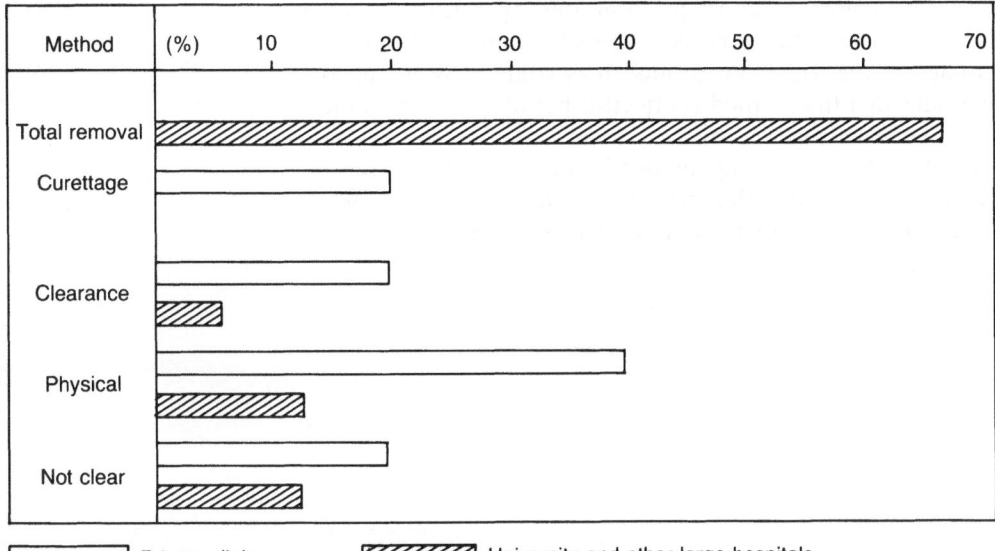

Fig. 14.15. The difference in treatment preferences between university hospitals and private clinics (1977). University hospitals overwhelmingly prefer the complete removal method, whereas private clinics prefer methods that yield superior aesthetic results. (Modified from Inaba 1986 with permission)

a recent period of 7 years show only 22 cases (only 3 cases per year) treated for bromidrosis (Nishimura et al. 1986). Another university reports only about ten patients in 1 year (Yano et al. 1988). This indicates that many patients eschew an operation in these large hospitals.

Aesthetic surgeons in private clinics prefer the clearance method (private clinics 20% vs university hospitals 6%) and the physical (electrocoagulation) method (private clinics 40% vs university hospitals 13%). The latter operations take much time and skill to remove sweat glands in full. Plastic surgeons favor taking such complicated procedures based on the satisfaction of the patients in not remaining with large scars.

Having met a great number of bromidrosis patients who had undergone various types of operations with no significant results, we keenly felt the difficulty of the radical operation for bromidrosis that had been performed in a series of trial and error procedures. However, it would be of no significance to perform the Skoog method, a radical operation which we consider to be not fully effective. The current technique is the suction method, but it would again be not recommended to perform this operation—at extremely high treatment costs—without conducting thorough studies of the follow-up results. It is completely wrong to consider that the method is effective because no complaints have been received from the patients after the operation. We should always keep in mind that the patients never come back because the surgical results are unsatisfactory and that they refuse to undergo another operation at the same clinic.

Chapter 15.   The Inaba Method

The authors have developed a radical technique for the treatment of bromidrosis and hyperhidrosis which utilizes a subcutaneous tissue shaver. For complete removal of sweat glands and hair follicles with only a small incision, an instrument combining a razor-type blade and a counter-pressure component was considered to be ideal. It took 4 years to produce a trial instrument, because manufacturers did not want to produce only one on the whim of a clinical physician.

15.1   Subcutaneous Shaver

The instrument consists of two jaws, one containing two notched rollers for counterpressure on the skin (Fig. 15.1). Each roller is 1 cm in diameter and is attached to the frame by an axis. The distance between the two roller axes is 1.5 cm and the length is also 1.5 cm (Fig. 15.2.a). The other jaw has a replaceable razor blade for shaving the subcutaneous layer from the undersurface of the skin. The blade width is 1.5 cm and the angle can be set at varying degrees (Fig. 15.2.b). With this instrument, the subcutaneous tissue of the axilla, including the sweat glands up to the level of a split skin graft, can be easily removed (Fig. 15.2.c).

The blade angle can be set for rough, medium, or complete shaving (Fig. 15.3). For rough shaving, the blade edge is tilted slightly away from the roller, and the blade is then introduced through the incision. Retraction on the chest causes tension on the skin. As a slight counterpressure is applied on the roller, the instrument is pulled backward to shave most of the subcutaneous tissue beneath the surface of the axillary skin. Almost all of the apocrine glands can be removed, leaving the eccrine glands and hair roots

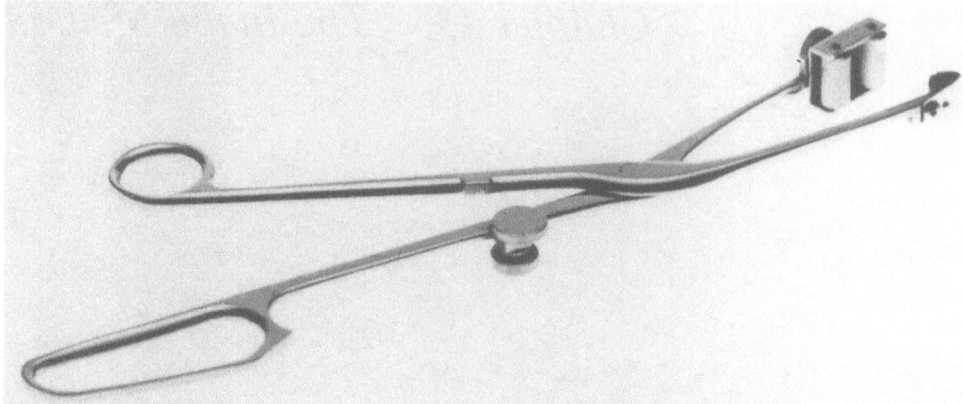

Fig. 15.1. The subcutaneous tissue shaver. This instrument, unlike an ordinary scalpel which leaves large scars, can be inserted beneath the skin to remove the sweat glands

a

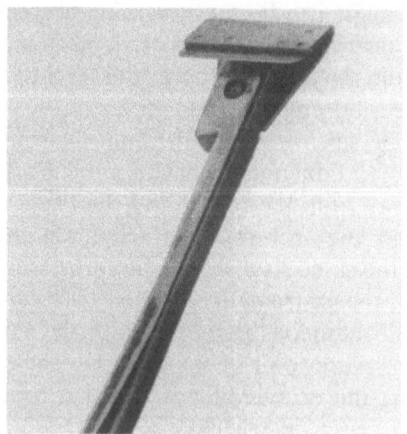

b

Fig. 15.2.a–c. The shaving instrument. **a** Roller portion, **b** Blade portion, **c** Setting position

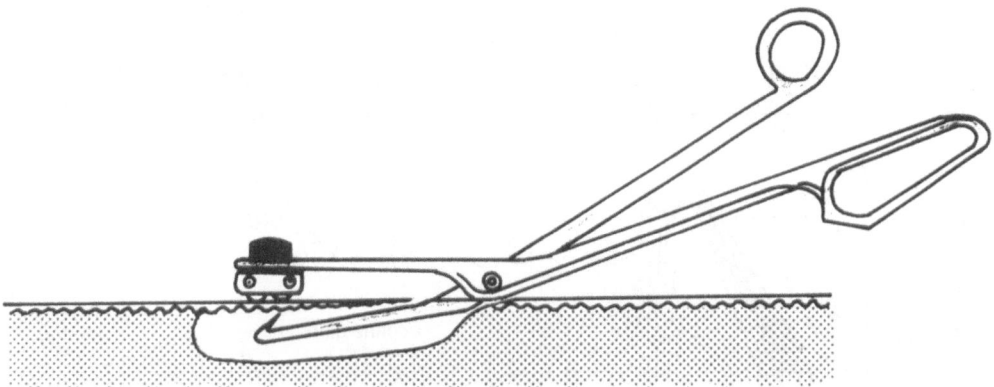

Fig. 15.2.c

Shaving Principle

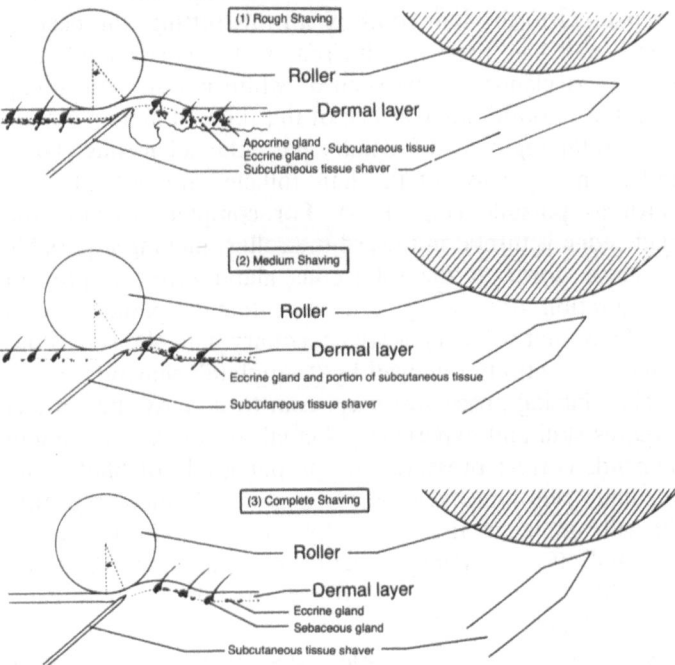

Fig. 15.3. Varying the angle of the blade in relation to the rollers leaves behind varying thicknesses of superficial skin. (Reproduced with permission from Inaba 1986)

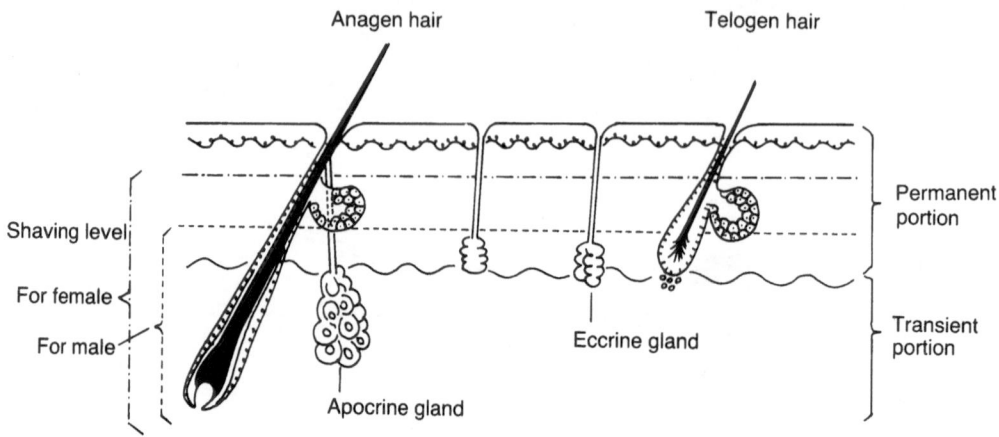

Fig. 15.4. Subcutaneous tissue shaving level. The shaving level for females prevents future regrowth of axillary hair, whereas a lower level setting leaves sebaceous glands and the upper isthmal portion of the follicles intact to ensure hair regrowth

intact. For medium shaving, the blade edge and roller are set parallel to the hair roots. With this setting, the eccrine glands and hair roots can be removed almost completely. Sebaceous glands can be seen as white nodules obtruding from the smooth undersurface of the skin. In male patients, who prefer regrowth of axillary hair, the sebaceous glands and isthmal portion of the hair follicles are left intact as much as possible (Fig. 15.4). For complete shaving the blade edge is turned in toward the roller, making it possible to shave down to the sebaceous gland layer to prevent regeneration of axillary hair. For female patients, who usually do not want regrowth of axillary hair, the sebaceous glands are entirely removed by "complete" shaving.

This shaving procedure may seem to be easy, but in fact requires skill and experience. Actual use of the instrument demands correct pressure, the proper grade of blade, and adequate skin tension. Pressure is gradually increased until the subcutaneous tissue is fully shaved. Skin tension is essential. If skin damage occurs, shaving cannot proceed smoothly. Shaving must be done with optimum pressure against the chest.

With the use of the complete shaving method up to the level of an intermediate split-thickness skin graft which removes the sebaceous glands, sweating and regrowth of axillary hair do not recur. For male patients who wish axillary hair regrowth with no sweating, the shaving must be done up to the level of a thick split-thickness skin graft, leaving the sebaceous gland (upper isthmal portion) intact.

15.2   Special Dressing

After shaving is finished, the skin flap must be fixed in place. Because the incision is small (1 cm), bleeding (hemorrhage) occurs, leading to hematoma and skin necrosis. To deal with this difficulty, a new type of dressing was developed. Called the "double tie-over dressing," it maintains hemostasis, fixation, and toleration of arm movement.

Although the subcutaneous tissue is shaved up to a medium split-thickness skin graft level, there is a difference between the skin shaved by a subcutaneous shaver and a common skin graft. The periphery of the shaved area can be considered a flap, since it is conjunctive with healthy skin, but the center of the shaved area has almost no blood supply and must be treated as a graft (flap to graft conversion). Since hemostasis is not performed and, furthermore, the axillary region is connected with a movable joint, a specific method of pressure fixation is necessary.

15.3   Skin Grafting

There are various depths at which different types of skin grafts are excised (Fig. 15.5). There are basically three types of grafts: epidermal, split thickness, and full thickness.

The full-thickness skin graft is a single large piece of skin which contains the epidermis and complete thickness of the dermis. Full-thickness skin is removed from the donor site by a scalpel. Its thickness varies with the skin thickness at the donor site, which is usually 0.32–0.40 mm in an adult. Fat tissue in the graft must be trimmed away from the base of the dermis. The fat tissue in the recipient area, however, is a barrier to the formation of new capillaries which grow into the graft from the host bed. If the subcutaneous tissue is not removed, revascularization does not occur from the fat tissue. Therefore, as much of the subcutaneous tissue must be removed as possible. A split-thickness graft contains the epidermis and a varied dermal thickness, ordinarily one-third to three-fourths. In this method subcutaneous tissue is not used, so revascularization and the final results are reliable.

The full-thickness graft is used to prevent contraction and retraction and to achieve less pigmentation, but is not

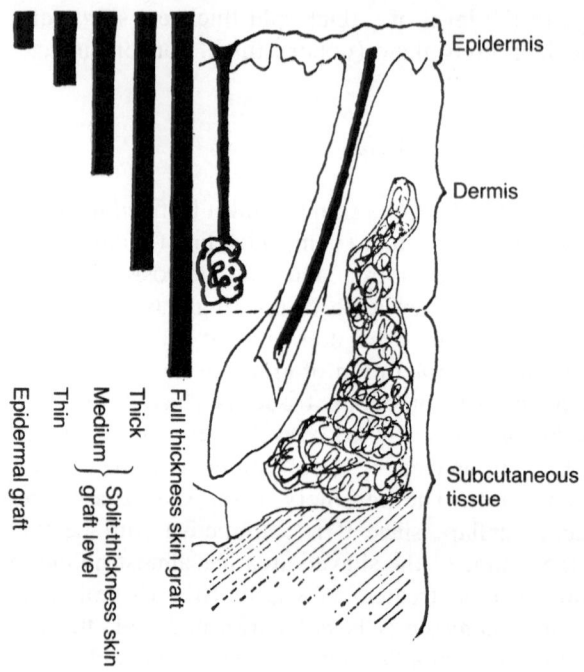

Fig. 15.5. Various levels of skin grafting. The different levels indicate depths at which different types of skin grafts are excised. The subcutaneous tissue must be shaved up to the level of a thick split-thickness skin graft. If the coiled duct (between the duct and secretory coil region) remains, sweat glands may be regenerated. (Reproduced with permission from Inaba 1986)

suited to bony and tendon sites of the body. In most cases, the donor skin is excised from the medial side of the thigh. For face grafting, however, the donor site is ordinarily behind the ear. For grafting in the region of the eye, the donor site is the supraclavicular region or the elbow. Donor-skin thickness is classified as thin, intermediate, or thick. A thin graft is usually 0.12–0.22 mm, and a thick graft is 0.22–0.30 mm. Split-thickness grafts are ruled out for the tendon, bony, and moving parts of the body.

These grafts (thin, thick, and split-thickness) are easier to handle than full-thickness grafts. The thinner the graft, the better the result. A complete epidermal graft is impossible. Epidermal grafts include both the epidermal layer and a thin layer of the dermis. Today these grafts are obtained with one of several types of electric dermatomes. It is difficult to obtain a uniform graft thickness with a manual dermatome or knife unless the surgeon can bring considerable skill and practice to bear.

The authors have used a patch grafting similar to an epidermal graft to an ulcerous area. A lidocaine injection is used for local anesthesia at the most superficial part of the skin between the dermis and epidermis before the razor is used for shaving. This epidermal grafting can be used on an area of inflammation and will decrease the inflammation rapidly. It is easier than other skin grafting, but its major drawbacks are contraction, retraction, and a pigmentation different from that of the graft area.

Conditions for successful transplantation include: (1) immobilization of the skin flap to ensure sufficient blood supply to the recipient site, (2) no gap between the skin flap and host bed, (3) no bleeding (hemostasis), (4) prevention of inflammation, (5) adequate skin expansion, and (6) no over-pressure.

After correct fixation with proper pressure on the skin flap and the underlying tissue, plasma circulation from the normal tissue to the flap occurs immediately. Two or three days after the shaving operation, blood circulation is observed. But if the flap fixation is inadequate, the flap will move, and circulation ceases. Grafts done on large deposits of fat, e.g., in extremely obese individuals, are at high risk because of the relative avascularity of this layer.

Flap circulation is inhibited in the presence of a hematoma, and causes flap necrosis. Bleeding must be avoided. If the size of hematoma is within 3 mm, however, it is no problem because of the so-called bridging phenomena, and necrosis does not occur. If the hematoma is removed within 3–4 days, the grafted skin can still survive. Sometimes, even 6 days after removal of the hematoma the skin flap can still survive. The worst complication is inflammation, which invariably causes necrosis. Another important factor is to avoid infection in order to obtain a successful graft. Preexisting skin eruptions such as seborrheic dermatitis often shelter many organisms and should be treated first.

15.4 Dressing

The following postoperative dressings have been used for skin grafts:
1. *The pressure fixation method.* The dressing is fixed with adhesive tape. The results are not good in areas where pressure cannot be applied sufficiently. Insufficient fixation causes hematoma, erosion, or necrosis.

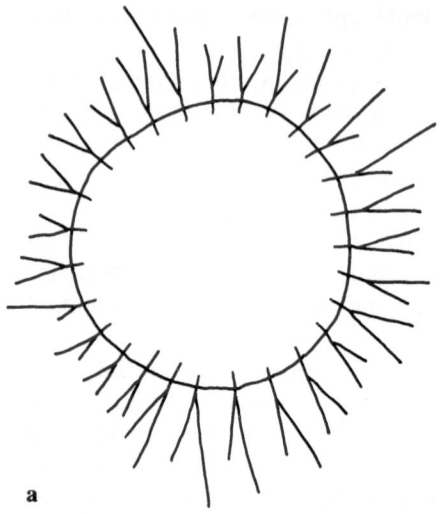

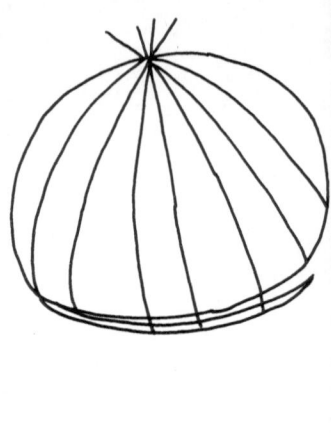

a   b

Fig. 15.6.a,b. The conventional single tie-over method. Sutures are placed along the edge of the graft and the ends are tied over the dressing

2. *The conventional single tie-over method* (Figs. 15.6.a,b). This was developed by Galtein in 1973. In this method, sutures are placed along the edges of the graft and the ends are tied over the dressing. This dressing is used mainly for pressure to obtain good adhesion between the recipient region and the graft since hemostasis is already done. However, hematoma is sometimes seen. The dressing is large and pressure is applied from the respective sides, which may cause the graft to shift. Even pressure cannot be applied over the whole graft when it is large, and thus hematoma can form. In this respect, Reese (1949) reported that the single tie-over method for skin grafts in the chest and abdomen has a risk of poor take.

3. *The reverse tie-over method*. This method was developed by Watanabe et al. (1967) and is used for a thick, free skin graft applicable to areas such as the extremities where the graft is larger than one-half the circumference of the extremity. The tie-over is done on the side opposite to the graft and folding or necrosis of the graft can be prevented.

To obtain a good adherent effect on the skin, some requirements must be satisfied. One is proper pressure and adequate immobilization. For that purpose, many

Fig. 15.7. The single tie-over method used in the axillary region

physicians use the conventional single tie-over dressing (Figs. 15.6.a,b) to tie the skin graft from the donor side to the recipient side, which is perforated by small incisions and then sutured. Saline solution is injected beneath the skin flap to clean any bleeding. Sofratulle, a gauze including fradiomycin sulfate (Roussel Laboratories, Tokyo), or other nonadherent material is used depending upon the size, and then the graft is tied up over the dressing gauze using the thread in suture between the graft and recipient region (Roussel Laboratories 1958). The difficult part of this procedure is the tie-up. If it is too strong, over-pressure occurs and causes necrosis. Under-pressure, on the other hand, results in hematoma.

15.5   Development of a Special Dressing

At first we used the single tie-over dressing to prevent hemostasis for the treatment of bromidrosis (Fig. 15.7). A thread was passed through the healthy skin outside the graft area superficially. Due to inadequate pressure, bleeding still occurred and hematoma and necrosis developed. The skin

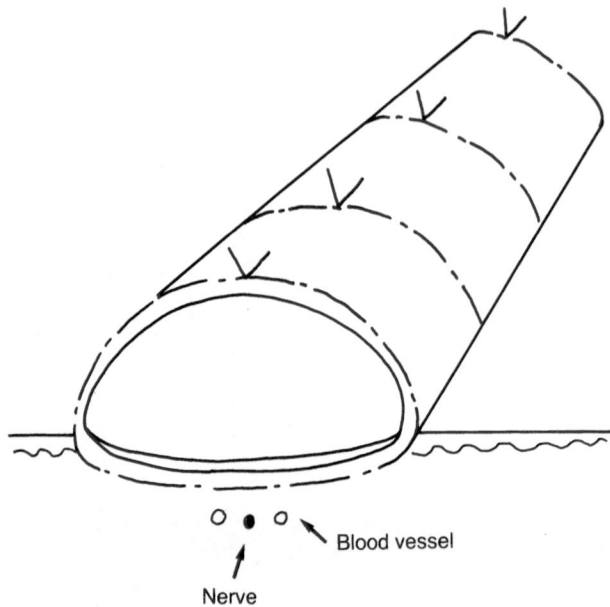

Fig. 15.8. The modified tie-over method

flap also moved to the center of the shaved skin and caused skin wrinkling because the dressing was too large. To counteract these drawbacks, we changed the suture technique (Inaba et al. 1978a). In the modified single tie-over method (Fig. 15.8) the suture is routed from the healthy skin into the tissue below the skin flap (not including the nerves and blood vessels), continued to the other side of the normal skin tissue, and tied up over the dressing. Even with this method it is difficult to control the pressure and remove the dressing. Sometimes this technique damages the nerves and blood vessels if the suture is too deep, especially in the axillary region. Single tie-over dressings can be used efficiently only in cases of nonimportant tissue, without nerves and blood vessels of the body surface.

15.6   New Method: "Double Tie-over Dressing Method"

The double tie-over dressing consists of 2 parts: dressing A and dressing B (Figs. 15.9.a,b). Contamination of dressing A with blood will harden the dressing causing over-pressure and necrosis. To minimize blood contamination, dressing

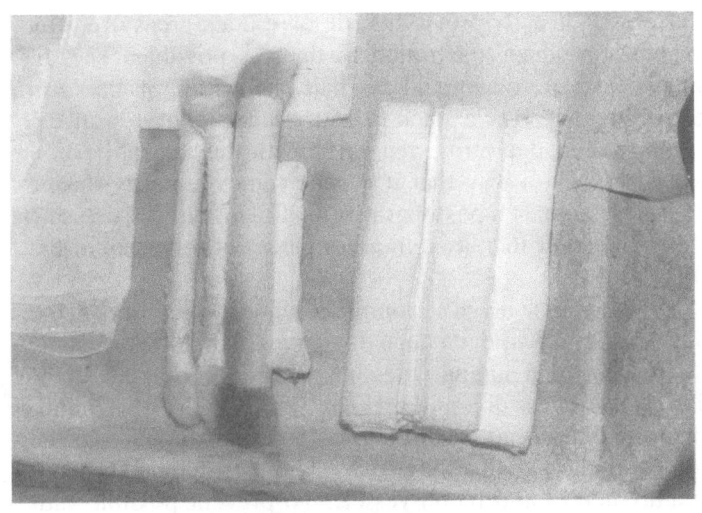

Dressing A

Dressing B

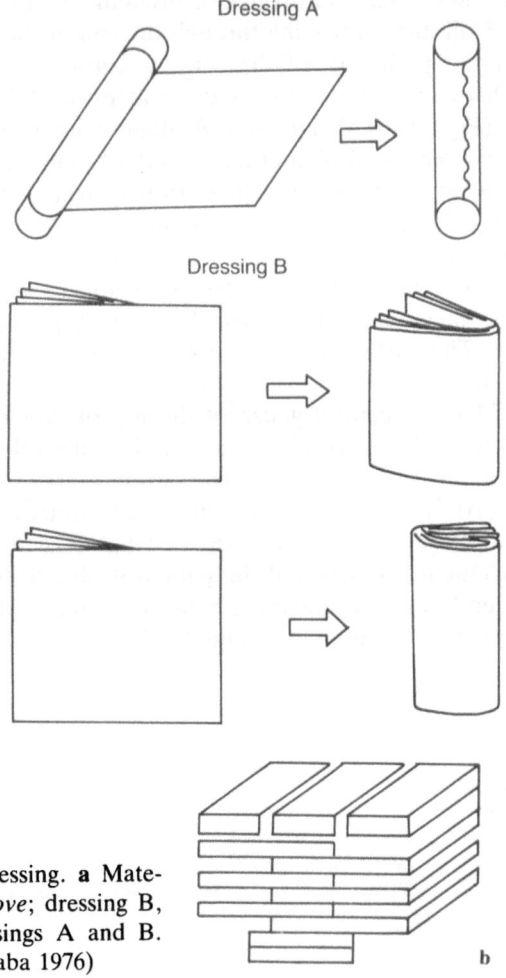

Fig. 15.9.a,b. The double tie-over dressing. **a** Materials of the dressing. Dressing A, *above*; dressing B, *below*. **b** Method of producing dressings A and B. (Reproduced with permission from Inaba 1976)

A consists of a strip of synthetic or crude (greasy) cotton wrapped in gauze and rolled as tight as possible. The diameter of the roll should be half the width of the skin graft. Dressing B is a stack of gauze layers. The top of the dressing is divided into three parts, with the central portion placed in such a way that it can be removed easily during decompression 24 h postoperatively. The volume of dressing B is reduced or increased in accordance with the size of the skin graft.

In order to achieve a complete hemostasis, control the pressure, and avoid damage to nerves and blood vessels, we developed a unique dressing which we call the "new double tie-over dressing" (Figs. 15.10.a–d). A round, cigarette-like form of dressing is first positioned on the center of the axilla, followed by careful stitching to avoid healthy nerves and blood vessels. To prevent possible sliding of the graft skin, both sides of dressing A are sutured with No.8 silk thread passing through the rim of the normal skin and the periphery of dressing A. Suturing is up to the width of dressing A and is done at every 1.5–2.0 cm intervals (Fig. 15.10.a). Dressing A alone is inadequate, so dressing B comprised of multiple layers of gauze is added (Fig. 15.10.b). The suture is tied strongly while lifting up so that dressing A is pressed firmly on the graft (complete hemostasis) (Fig. 15.10.c). Compression of the dressing is achieved when both lateral gauzes of the top of dressing B are slid to their respective sides and strongly pressed with the palm. The surface of the dressing is thereby flattened (Fig. 15.10.d).

After 24 h, the central gauze of the top of dressing B is removed to facilitate blood circulation. Two days thereafter further layers are removed, and suture marks are also reduced. After 6 days all dressings and sutures are removed, and the grafted site is dressed lightly with gauze. Arm movement is easier and the patient is able to function independently without nursing assistance. After 10 days the patient can perform all movements freely.

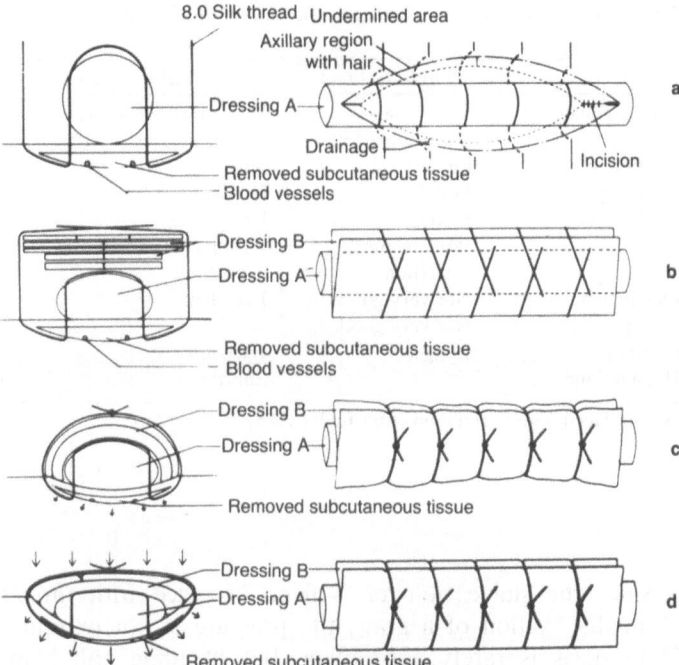

Fig. 15.10.a–d. The new double tie-over method. This dressing was developed for postoperative application after subcutaneous shaving for bromidrosis and hyperhidrosis and excellent results were obtained. The dressing is smaller in volume and shifting of grafted skin does not occur with motion. Blood circulation in the graft improves and the suture mark is also reduced by decompression of dressing B 24 h later. (Reproduced with permission from Inaba 1985)

15.7 Comparison Between Single Tie-over and Double Tie-over Dressings

The double tie-over dressing has been demonstrated as superior in all respects (Table 15.1):

1. The fixation and pressure of the double tie-over dressing are better than that of the single tie-over.
2. The material used in the double tie-over is less than that of a single tie-over.
3. Fixation of the shaved skin flap with the double tie-over is superior to that of the single tie-over.
4. Healing time is shorter with the double tie-over.
5. In order to obtain complete hemostasis, tie-up of the thread must be strong (tight); for that reason, visible suture marks remain, but do fade away over time.

Table 15.1. Comparison between the single and double tie-over dressings

	Single tie-over	Double tie-over
Dressing material	Cotton, gauze, chemical cotton	Synthetic cotton (dressing A) Gauze (dressing B)
Amount	Greater	Smaller
Absorption	Good	Excellent
Elasticity	Good	Excellent
Pressure	Weak central portion	Strong central portion
Evacuation caused by pressure	Not very good Not very good	Excellent
Fixation	Longer	Excellent
Healing time		Shorter

(Reproduced with permission from Inaba 1986)

Since the single tie-over is fixed in place with thread as in the fixation of a graft, the pressure is not extreme. Hemostasis is rarely a problem, but bleeding and hematomas sometimes occur due to lack of extreme pressure and less evacuation of retained blood. With the double tie-over dressing, especially dressing A which is fixed (sutured) strongly in accordance with its width extending from the healthy skin, evacuation from the recipient region is excellent.

The most characteristic advantage of the double tie-over dressing is averaged control of pressure postoperatively to obtain sufficient blood circulation for the shaved skin. Single tie-over dressings cannot provide for controlled removal of the dressing. Furthermore, with single tie-over dressings, the dressing is large and projects from the skin surface. In contrast, the surface of a double tie-over dressing is flat, due to regulating dressings A and B relative to the skin surface. Excellent fixation is obtained even in a moving area, such as the axillary region.

15.8   Surgical Procedure

15.8.1   Preparation Before Operation and Anesthesia

Figure 15.11 shows the instruments and materials necessary for the subcutaneous tissue-shaving procedure. The physician cannot measure the blood pressure in a conventional

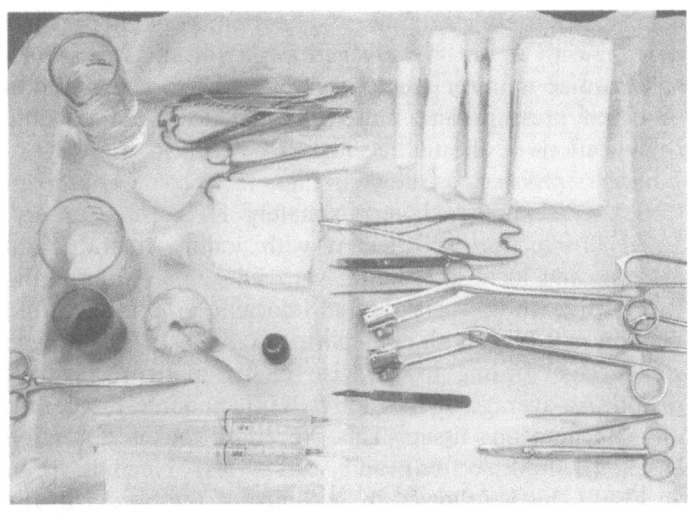

Fig. 15.11. Layout of surgical requirements

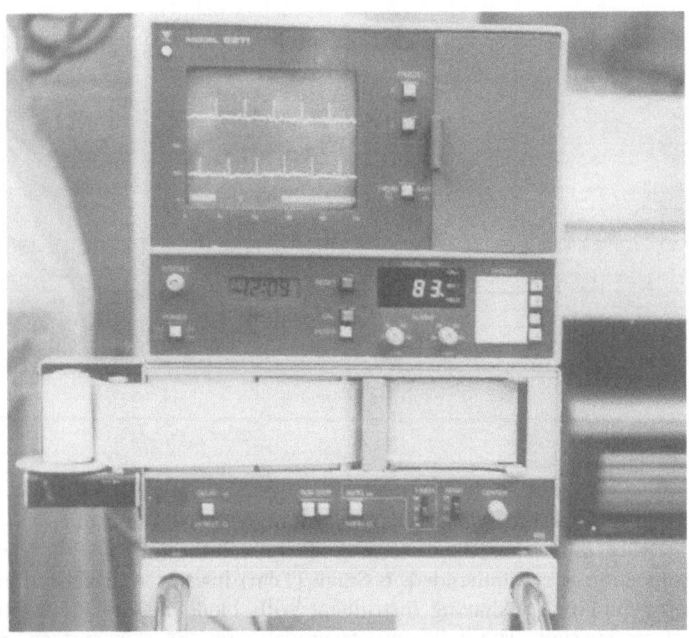

Fig. 15.12. Blood pressure monitored on the leg

manner because the arm must be held up. A blood pressure monitor with an air bag attached to the lower limbs as well as a cardiac monitor must be employed for due attention to blood pressure and pulse in order to avoid possible complications of anesthetic shock (Fig. 15.12).

Surgery proceeds as shown in Figs. 15.13.a–m. Diazepam (5 mg) is administered approximately 1 h before surgery. Local disinfection is affected with iodine tincture and alcohol, and local anesthesia, with lidocaine. Considering its toxicity, we use very diluted lidocaine (1% in eightfold dilution = 0.14%) with epinephrine (1:100000). The dosage is varied according to the size of the patient's axillary region. An average of 100–150 cc is injected into the dermis and subcutaneous tissue. This procedure makes dissection easier (hydrodissection) and causes less bleeding (Fig. 15.13.a). This is followed by outlining of the border of the axillary hair-bearing region with methylene blue. In order to achieve complete removal, the borderline must be drawn about 0.5–1.0 cm beyond the axillary region. A 1-cm incision is made at the distal part of the drawing on the upper arm (at the border of the hair-bearing axillary region) lengthwise along the upper arm (Fig. 15.13.b).

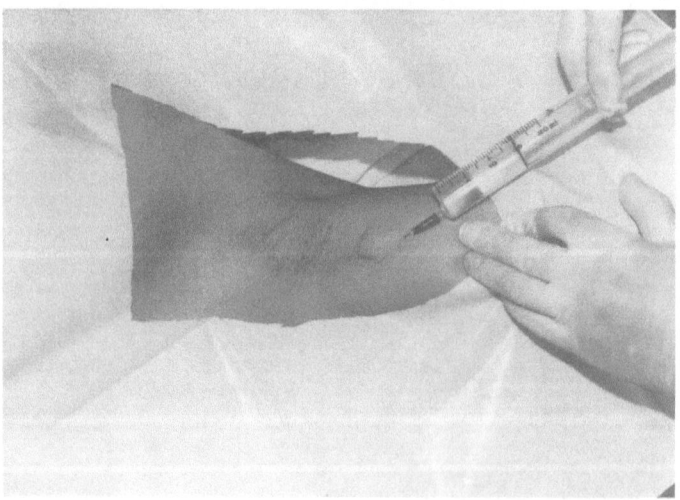

a

Fig. 15.13.a–d. The surgical procedure. **a** Local anesthetic using a very dilute anesthetic formula (0.14% lidocaine). A shallow injection is recommended. **b** Small (1 cm) incision and dissection by snipping. **c** Shaving instrument with blade inserted through 1-cm incision. The roller exerts pressure on the skin surface. **d** Complete shaving for female patients; axillary hair and sweat will not reoccur. (*Contd.*)

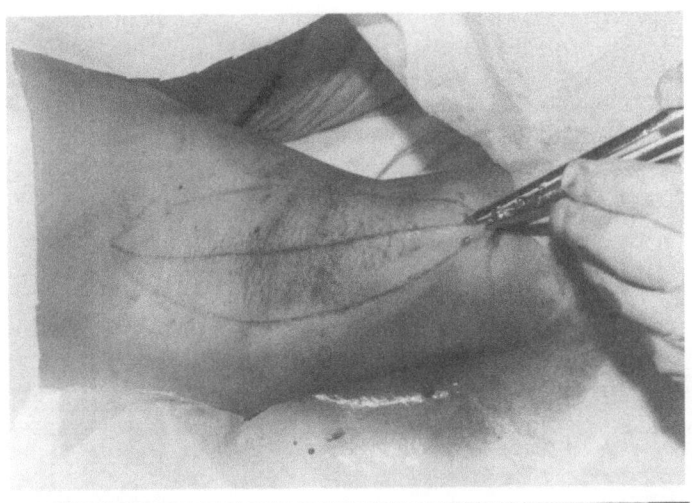

b

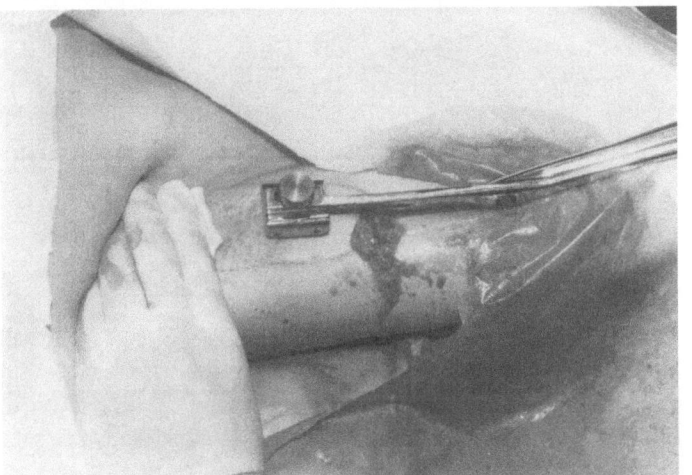

c

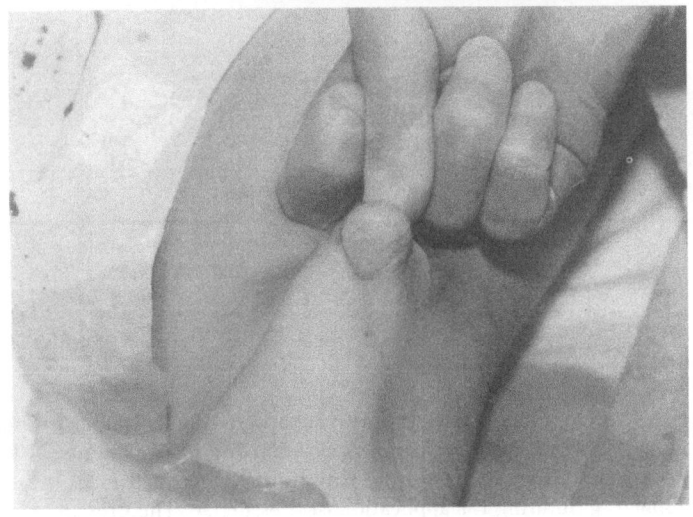

d

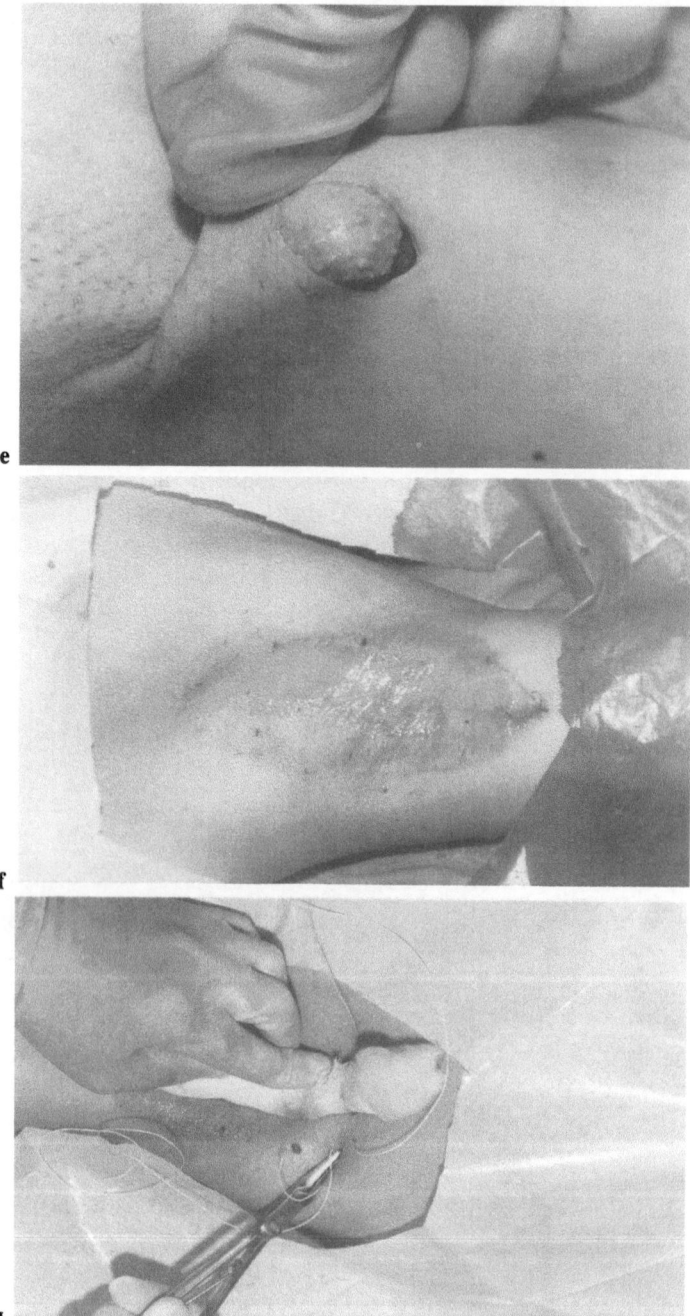

Fig. 15.13. e Medium shaving with sebaceous glands (upper isthmal portion of hair follicle) left intact to assure axillary hair regeneration for male patients. **f** This shaving method does not prevent bleeding when suturing is performed. A small incision is made and antibiotic ointment applied. **g** Suturing for application of dressing A. The dressing is positioned along the border of

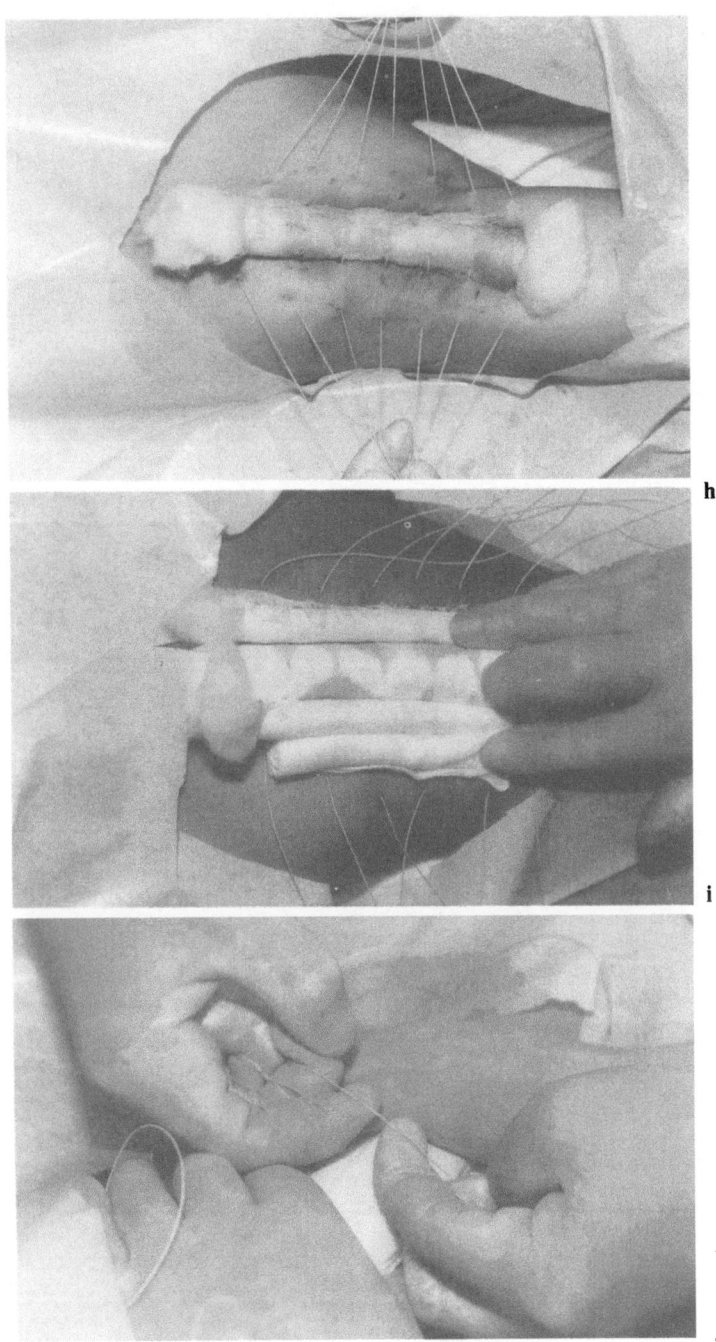

the pectoralis major, not at the central area. **h** Appearance of deep suturing. **i** A small dressing A is added, depending on the width of the shaved skin. **j** Dressing B is placed over dressing A. The thread is pulled strongly and tied. (*Contd.*)

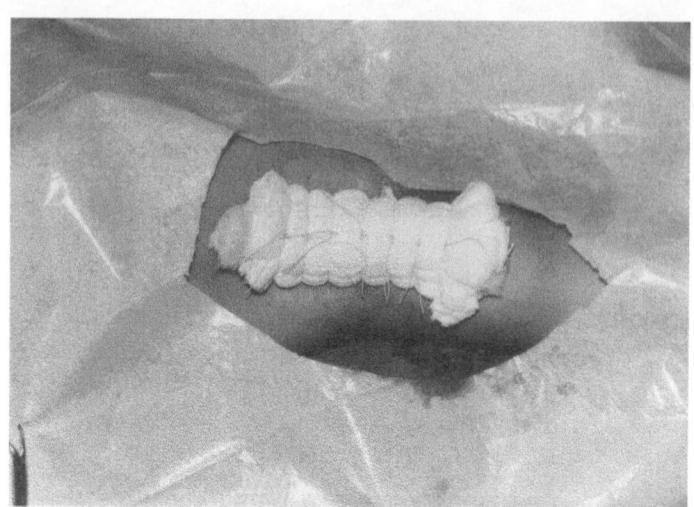

k

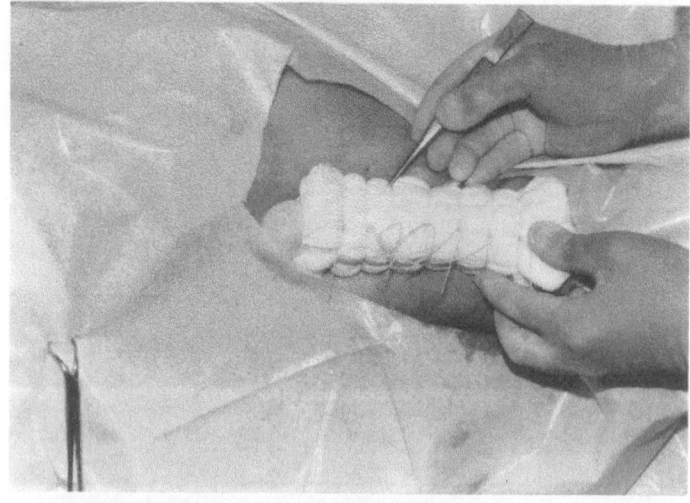

l

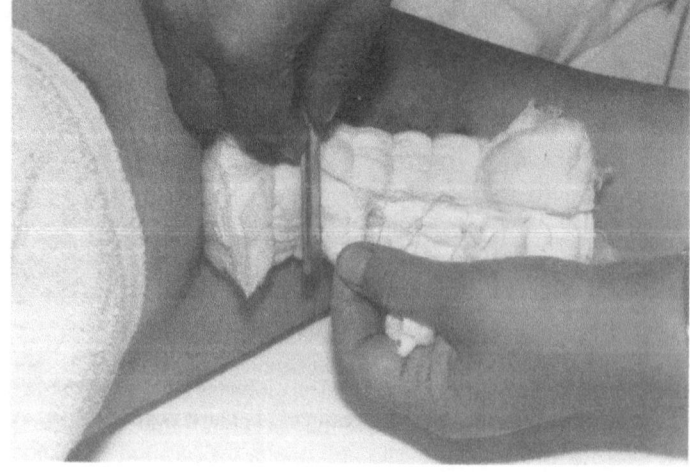

m

15.8.2   Operation Method

Dissection by snipping with scissors follows the outlining. The complete or medium shaver is introduced through the small incision, parallel to the skin surface. Constant pressure moves the shaver forward and back with gentle but fluid motion. Both speed and pressure of shaving are important (Fig. 15.13.c).

With complete shaving (up to medium split-thickness skin graft) for female patients, axillary hair and sweat will not recur (Figs. 15.4, 15.13.d, 15.14.a–e). With medium shaving (up to thick split-thickness skin graft) for male patients, sebaceous glands (upper isthmal portion of hair follicle) are left intact to assure axillary hair regeneration (Figs. 15.4, 15.15a–e).

This shaving method does not prevent bleeding, so a complete dressing is needed. Adherent polyp-like fatty tissue at the margin must be removed (Figs. 15.18.a,b). These tissues can be a source of interference with the pressure and thus cause hematoma. We use gauze to slowly clean the shaved tissue. The polyp-like fatty tissue can then be pulled out with the gauze. A small incision is made on the shaved skin graft for evacuation of blood.

Suturing is performed after confirming a reduction of bleeding. Dermal suturing at the incision site is done to prevent the occurrence of a keloid. If bleeding is excessive, we use gauze to ascertain the bleeding area. A small incision in most cases of bleeding is made in the skin graft to confirm that bleeding has stopped. After suturing the incision, the shaved skin is pressed again strongly to evacuate remaining blood and to reexamine any polyp-like fatty tissue at the small incision site. Before affixing dressing A, an antibiotic wax (chloramphenicol wax) ointment is rubbed on the shaved skin (Fig. 15.13.f).

Fig. 15.13. k Appearance of double tie-over dressing when completed. **l** Dressing B modified for maximum application. **m** Partial removal of dressing B after 24 h, to abet blood circulation

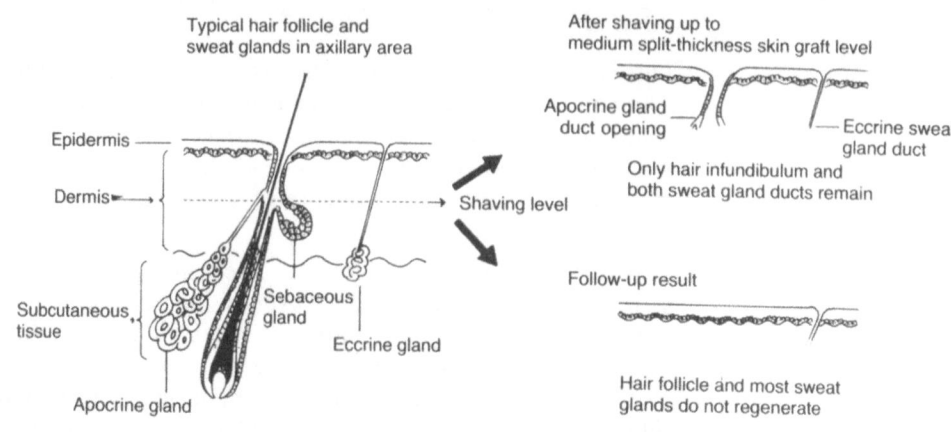

Typical hair follicle and
sweat glands in axillary area

After shaving up to
medium split-thickness skin graft level

Epidermis

Apocrine gland
duct opening

Eccrine sweat
gland duct

Dermis

Shaving level

Only hair infundibulum and
both sweat gland ducts remain

Subcutaneous
tissue

Sebaceous
gland

Eccrine gland

Follow-up result

Apocrine gland

Hair follicle and most sweat
glands do not regenerate

a

After treatment

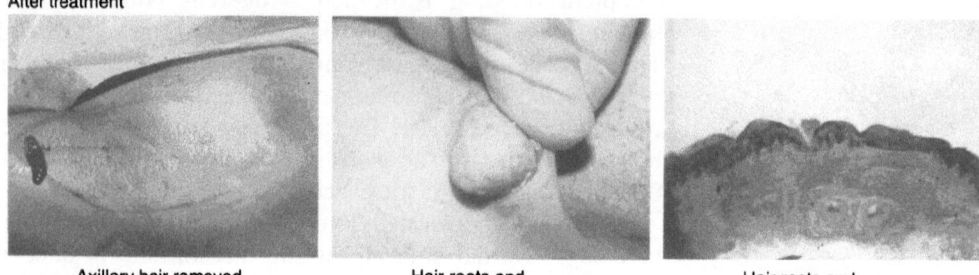

Axillary hair removed

Hair roots and
sebaceous glands
almost removed

Hair roots and
sweat glands
completely removed

Postoperative follow-up results

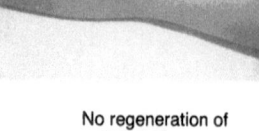

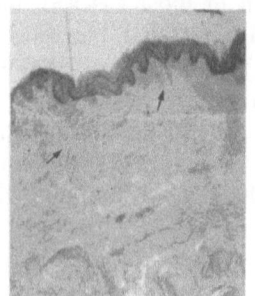

No regeneration of
axillary hair
after 2 years

No regeneration: only
a few eccrine ducts
remain

Only upper portions
of eccrine ducts
remain

b

Fig. 15.14.a–e. Subcutaneous tissue shaving up to medium
split-thickness skin graft level. **a** Diagrammatic representation of shaving level and postoperative results. **b** After treatment. **c** Female patient after complete subcutaneous tissue
shaving. No axillary hair or sweating appear. **d** Postoperative
results. **e** Postoperative results

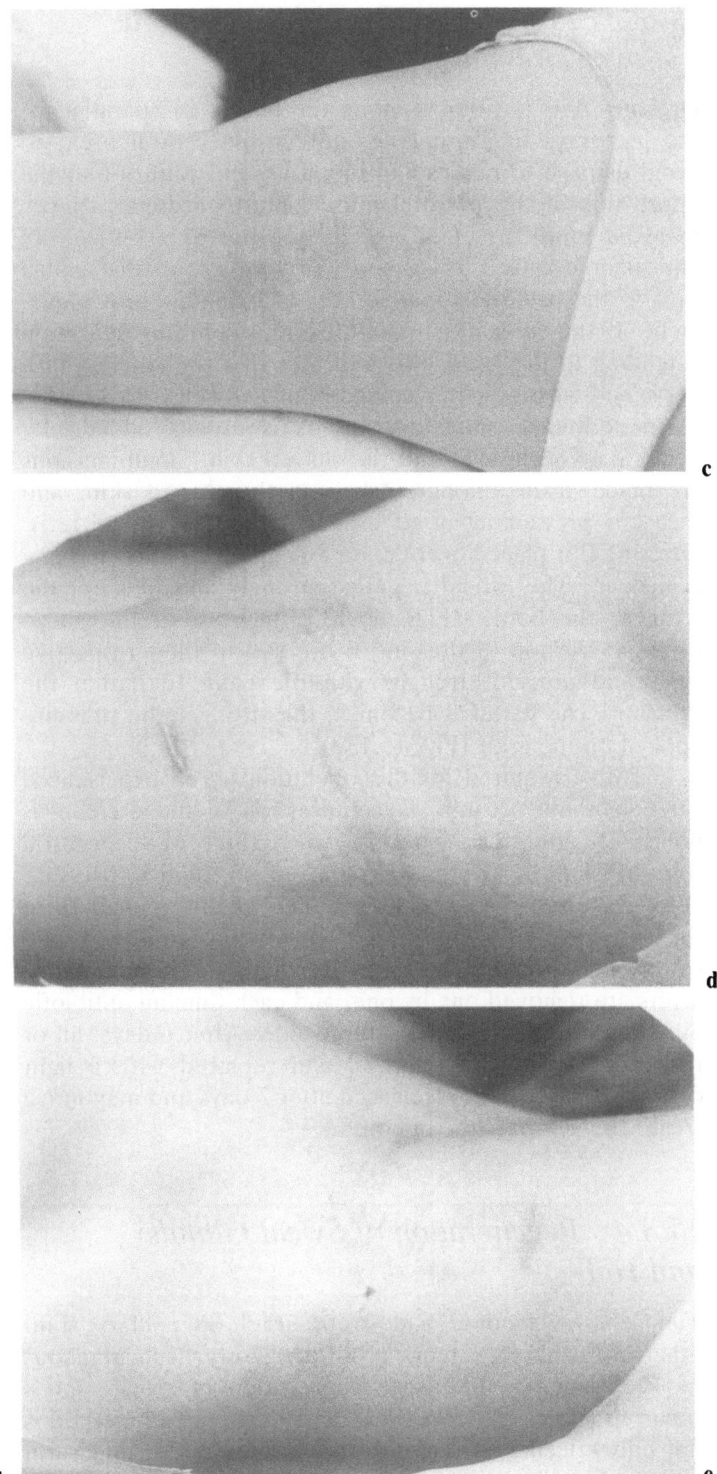

Fig. 15.14.

15.8.3   Postshaving Procedure (Double Tie-over Dressing)

Dressing A is positioned along the border of (parallel to) the pectoralis major muscle, not at the central area, to avoid damage to nerves and blood vessels. Suturing at the upper side of the pectoral muscle must be done expertly to avoid injury to blood vessels and nerves. This is very important because nerves and blood vessels in the axillae shift to the pectoralis muscle area in extended arm movement. If the suturing is painful, e.g., a burning sensation extending to the hand, this indicates that the stitch is in a nerve and must be removed and relocated (Figs. 15.13.g,h).

An additional small dressing A is usually added, depending upon the width of the shaved skin. Small incisions are made again among stitches on the shaved skin, and the skin pressed again to evacuate blood (Fig. 15.13.i). Dressing B is placed over dressing A and the entire dressing is tied up. The thread is pulled strongly and tied up, the stronger the better (Figs. 15.13.j,k). Both of the gauze layers at the top of dressing B are slid to their respective sides and pressed strongly with the palm to flatten the dressing. The flatter it becomes, the stronger the pressure applied on the graft (Fig. 15.13.l).

The time required for this operation by an experienced surgeon is only 30 min. Decompression of the dressing is done 24 h after the operation, when three of the central top dressing B layers are removed to reduce pressure (Fig. 15.13.m). The pressure is then gently applied from the lateral sides, with blood circulation improved and suture marks reduced. Each day thereafter, the gauze layers are removed one by one, and each time an antibiotic ointment is applied to the sutured side. After 6 days, all of the gauze layers are removed and replaced with a light dressing. The patient is released after 3 days and may move freely 10 days after the operation.

15.8.4   Regeneration of Sweat Glands and Hair

The removal method had to be used on axillary skin, otherwise complete removal of the lower subcutaneous tissue was impossible, and regeneration of sweat glands occurred as a result. However, even this removal method is not fully effective in removing the entire area of underarm hair, making it an incomplete removal method. Now, with the development of the subcutaneous shaving method, the skin can be completely shaved up to the medium split-thickness skin graft level, and no regeneration of sweat glands occurs. This method has solved the problem of hyperhidrosis.

15.8.5 Shaving Depth for Effective Removal

We now discuss the level at which the subcutaneous tissue should be shaved from the reverse side of the skin in order to prevent regeneration of sweat glands (Inaba et al. 1990a). If regeneration of the sweat glands occurs from the lower tip of the remaining duct, methods of removal from beneath the skin may be ineffective in the long run. A decision must be made on how much shaving should be done for appropriate removal.

Shaving up to Medium Split-thickness Graft Level. If shaving is done up to the medium split-thickness graft level, no regeneration of hair and sweat glands will occur, even though the ducts of apocrine and eccrine glands remain (Figs. 15.14.a–e). Follow-up results in both clinical and histological research support this conclusion. In turn, postoperative follow-up results show no regeneration of axillary hair after 2 years. Histological findings show that only a few eccrine ducts can be seen; the pilosebaceous unit and apocrine glands have disappeared completely. An enlargement of the view reveals atrophy of the eccrine duct.

Shaving up to Thick Split-thickness Graft Level. If shaving is set for a thick split-thickness graft, the sebaceous glands and the ducts of both sweat glands remain (Figs. 15.15.a–e). The sebaceous glands can be seen in a protrusive condition without surrounding connective tissue. Axillary hair after shaving remains by retention of the upper inner root sheath even if most of the lower hair follicle is removed. In follow-up results, hair regeneration is observed, but the sweat glands do not regenerate. We reported (Inaba 1986) that the subcutaneous tissue shaving method revealed that regeneration of hair can occur from the upper isthmal portion of the follicle adjacent to the duct opening of the sebaceous gland even if telogen hair bulbs have been removed. However, shaving at a level below the isthmal portion results in no regeneration of apocrine and eccrine glands even if their ducts remain.

Shaving up to Full-thickness Graft Level. With shaving up to the full-thickness graft level, almost all connective tissue

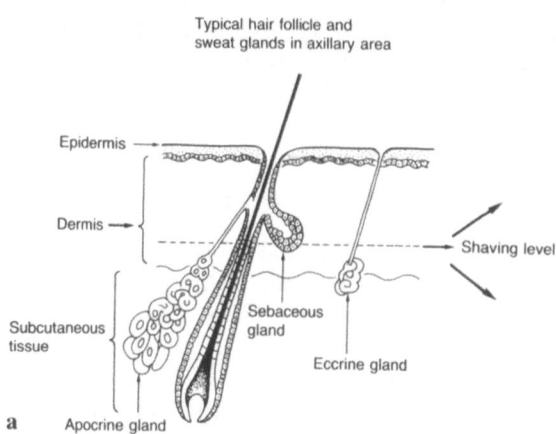

Typical hair follicle and
sweat glands in axillary area

Epidermis

Dermis

Shaving level

Subcutaneous
tissue

Sebaceous
gland

Eccrine gland

a   Apocrine gland

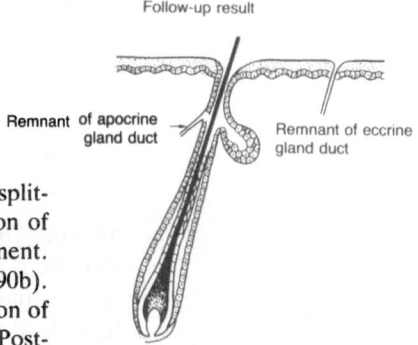

After shaving up to
thick split-thickness skin graft level

Apocrine gland
duct remains

Sebaceous
gland

Eccrine sweat
gland duct remains

Inner root sheath

**Inner root sheath remains, so hair is preserve
even though the lower portion of the follicle
and sweat glands coil have been removed.**

Follow-up result

Remnant of apocrine
gland duct

Remnant of eccrine
gland duct

Fig. 15.15.a−e. Subcutaneous tissue shaving up to thick split-
thickness skin graft level. **a** Diagrammatic representation of
shaving level and postoperative results. **b** After treatment.
(Reproduced with permission from Inaba and Inaba 1990b).
c Male patient after thick shaving. In spite of regeneration of
axillary hair, no odor and only scant sweating occur. **d** Postoperative results. **e** Postoperative results

**Hair follicle fully regenerates:
sweat glands do not regenerate**

After treatment

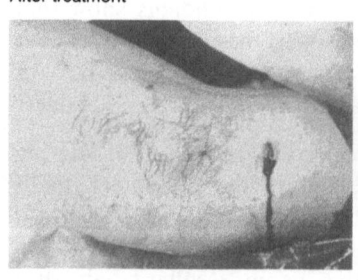

Axillary hair remains
after shaving

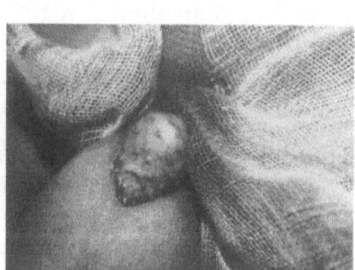

Sebaceous glands
visibly protrude;
lower portion of
follicle removed

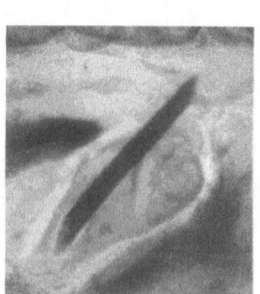

Sebaceous gland
and upper (isthmal)
portion of follicle remain

Postoperative follow-up results

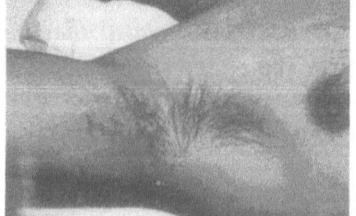

Hair regrowth observed 6 years later

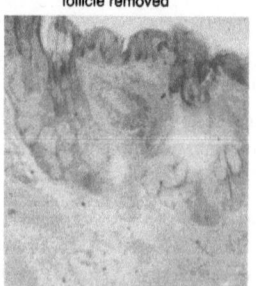

Sweat glands have not been
regenerated; hair regeneration
has occurred from upper
isthmal portion of follicle

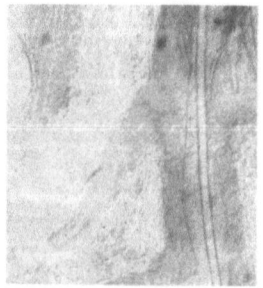

Apocrine duct atrophied

b

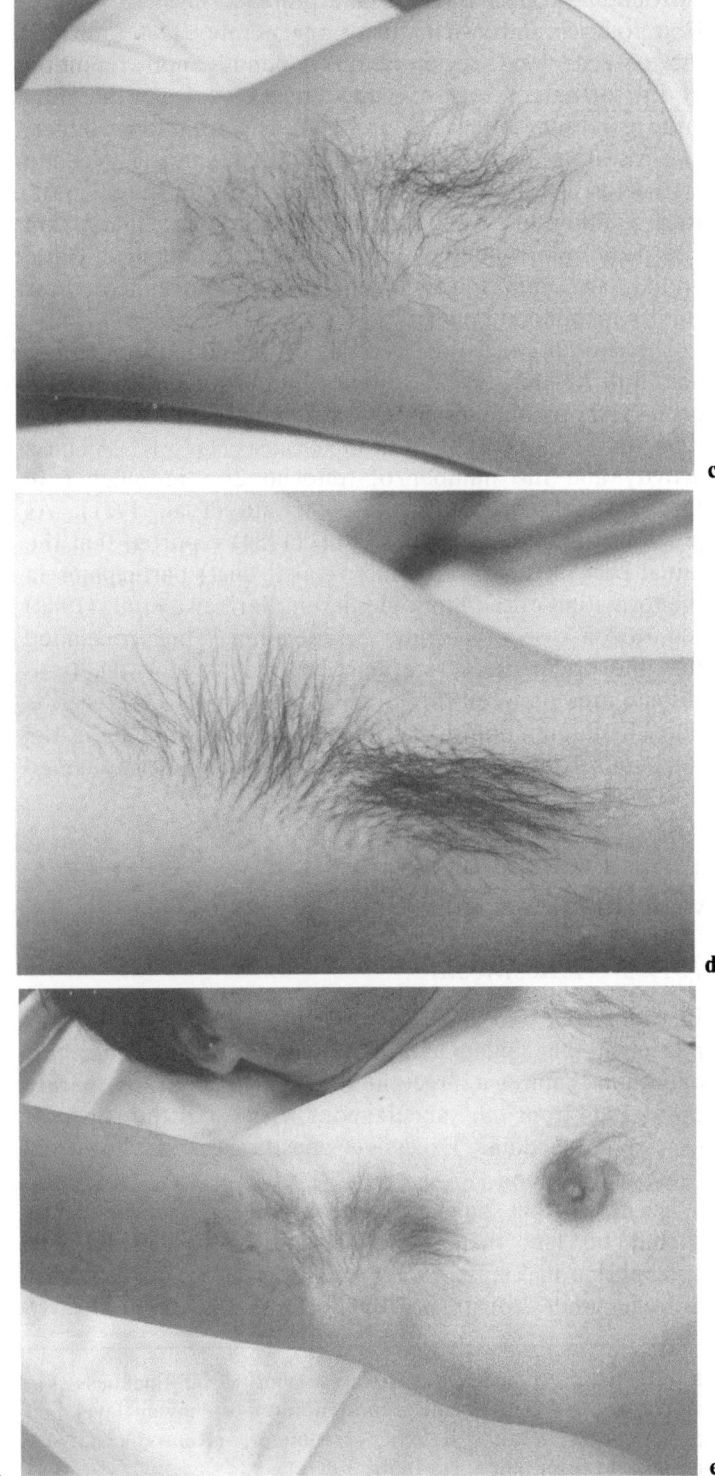

Fig. 15.15.

surrounding the sebaceous gland remains (Figs. 15.16.a,b). Hair follicles and sweat glands regenerate. This indicates that sweat gland regeneration depends upon retention of the boundary region (coiled duct) between the duct and secretory coil. The problem of bromidrosis/hyperhidrosis does not recur, provided that shaving is done up to the medium level of a thick split-thickness graft. Regeneration of sweat glands beneath the axillary skin depends upon whether the coiled duct remains. These findings are supported by biochemical and ultramicroscopic studies mentioned below.

The central sweat duct consists of an outer ring of peripheral or basal cells and an inner ring of luminal or cuticle cells. The proximal (coiled) duct is apparently more active than the distal straight portion because Na^+, K^+-ATPase activity and the number of mitochondria are higher in the proximal portion (Ellis 1967; Sato et al. 1971). As previously stated, Kurosumi et al. (1984) reported that the initial part of the dermal duct (coiled duct) participates in the formation of secretory glomeruli. Serikawa et al. (1988) reported a case of eccrine spiradenoma. They concluded that the origin of eccrine spiradenoma may be the transitional area between the secretory portion and the eccrine duct. Differentiation in eccrine spiradenoma may be in the direction of both the myoepithelial cell and eccrine coiled duct.

15.9 Problems Encountered with this Procedure

15.9.1 Anesthesia

In order to avoid damage to blood vessels and nervous plexus in the axillae and to obtain easy dissection, this operation requires a large amount of anesthesia to separate the dermis from the subcutaneous tissue and thus prevent excessive bleeding. Excessive anesthesia can, however, cause toxification.

Even with this shaving procedure, the maximum dosage should be less than 1% lidocaine hydrochloride with epinephrine in 5 mg/kg body weight or about 50 cc for the average adult. Most surgeons believe that the effective

→

Fig. 15.16.a,b. Subcutaneous tissue shaving to full-thickness skin graft level. **a** Diagrammatic representation of shaving level and postoperative results. **b** After treatment. (Reproduced with permission from Inaba and Inaba 1990b)

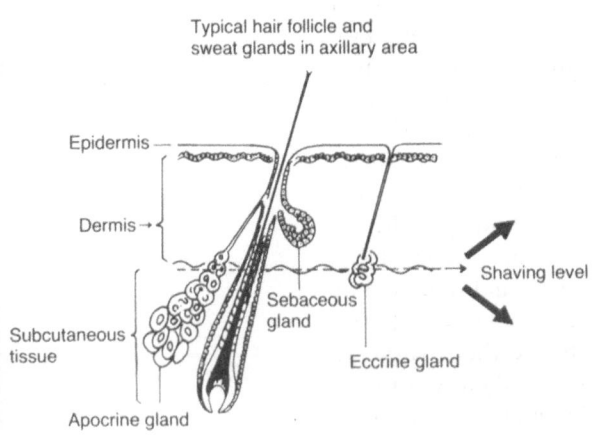

Typical hair follicle and
sweat glands in axillary area

Epidermis

Dermis →

Subcutaneous
tissue

Sebaceous
gland

Eccrine gland

Apocrine gland

Shaving level

After shaving up to
full-thickness skin graft level

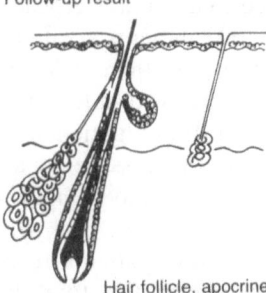

Connective tissue surrounding sebaceous
gland remains, as do upper portions of
apocrine and eccrine glands

Follow-up result

Hair follicle, apocrine gland and
eccrine gland are fully regenerated   **a**

After treatment

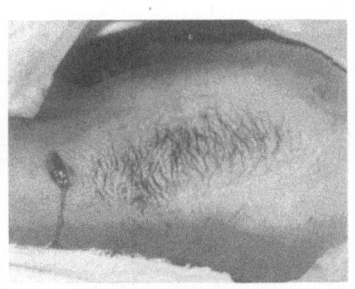

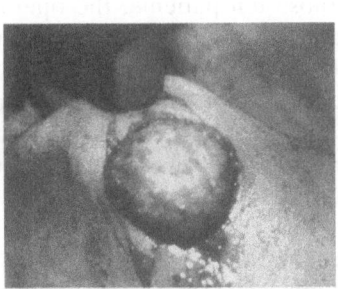

(a) Axillary hair remains

(b) Connective tissue surrounding
sebaceous gland

(c) Eccrine and apocrine
glands remain

Postoperative follow-up results

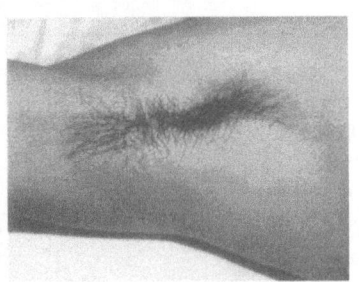

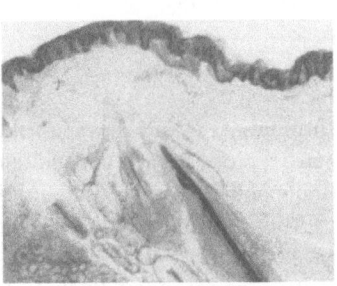

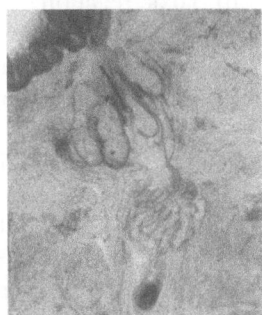

(a) Almost full hair regrowth observed
5 years later, slight odor observed

(b) Full regeneration of hair follicle
and both sweat glands observed

(c) Full regeneration of apocrine
gland and hair   **b**

duration of lidocaine (0.5%–1%) is 1–1.5 h and that for procaine (0.5%–1%) is 1 h (maximum dosage 10 mg/kg). A study of how much dilution of anesthesia is efficacious is shown in Fig. 15.17 (Inaba et al. 1978d, 1983). About 44% of all patients given this dosage required no oral pain-killing relief postoperatively. The effect of this local anesthesia continues for an average of 10 h. Diluted lidocaine has the same effect as an absolute dosage. This finding indicates that the dilute solution is locally efficacious. Klein (1986) used 0.1% lidocaine in liposuction surgery and obtained an efficacious effect with no side effects. Iwayama et al. (1988) also reported that 0.1% is efficacious for aesthetic surgery.

Of the 20,000 patients treated thus far, slightly lowered blood pressure (less than 20% of individual average blood pressure) was observed in less than 1%, about 200 patients. Observation of these patients was continued until there was a decrease in pressure to about 80 mmHg in systolic value, using oxygen and a slight hypertensive drug, e.g., a carnigen (Hoechst Co., Tokyo) or ephedrine injection. Almost all of these patients recovered immediately. Only ten patients among the total number developed severe shock. Six patients, with blood pressures below 70 mmHg in systolic value, responded well when a 0.2–0.3 cc adrenaline or a 0.5–1.0 cc ephedrine and corticosteroid injection was used. In four among those ten patients, the operation had to be stopped. An attempt to operate again was made the next day, with due consideration of the shock factor and use of a hypertensive drug or some other medication. If patients are in psychosomatic shock, too much medication is inadvisable, due to the risk of iatrogenic complications.

Complications of local anesthetics include:

1. Systemic reaction to local anesthetics
   a) Allergic reaction (anaphylactic shock or other allergic reactions)
   b) Toxic reaction
2. Systemic reaction to vasoconstrictors
3. Regional complications
4. Psychomotor reactions

Reaction anaphylaxis is immediate, and severe allergic response is characterized by rapid loss of consciousness, respiratory difficulty (bronchospasm), cyanosis, and no recordable blood pressure. Other allergic reactions (delayed reactions) include various combinations of cutaneous manifestations such as pruritus, generalized flush, edema, and urticaria. Toxic reactions result in light-headedness,

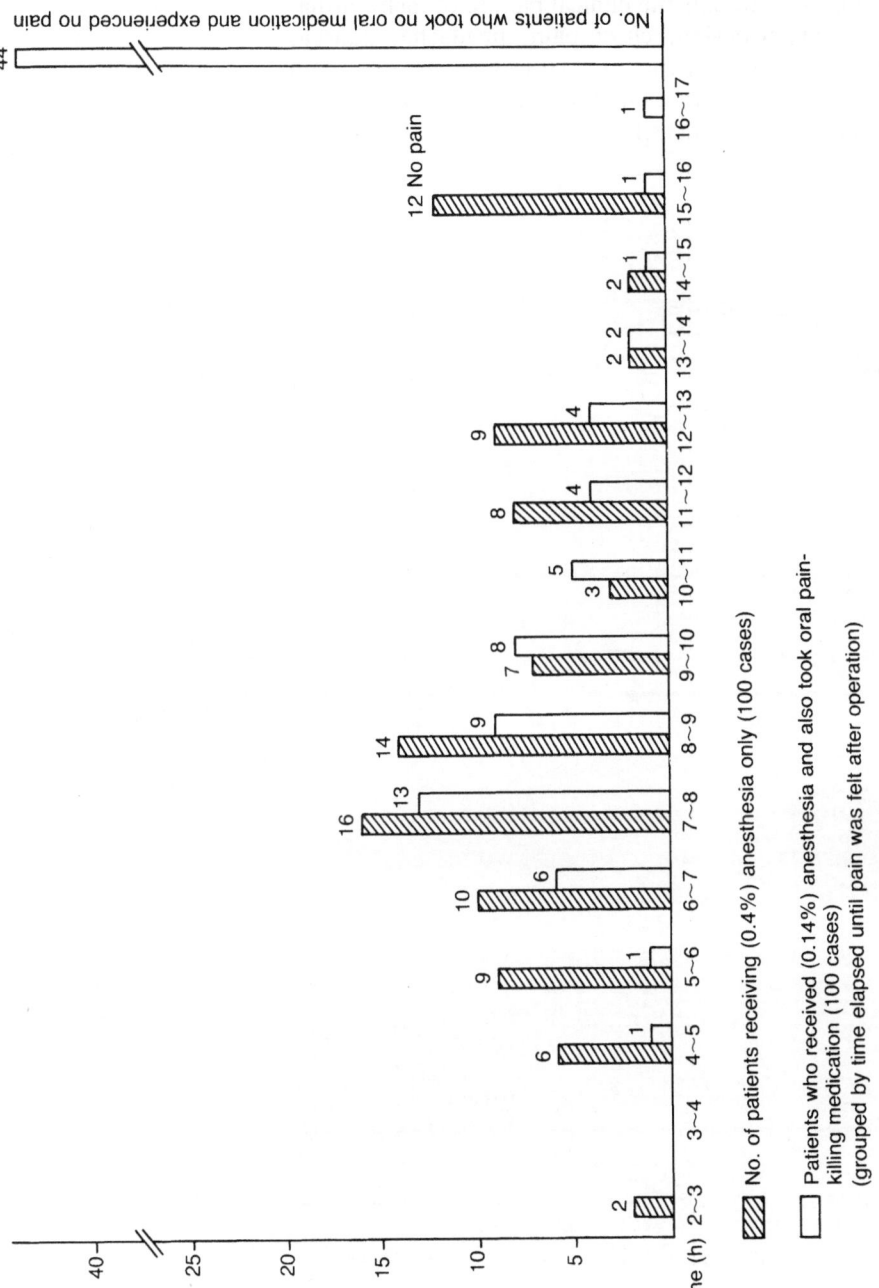

Fig. 15.17. Time which elapsed until pain began following operation

numbness, convulsion coma, and central nervous system depression. With regard to vasoconstrictor reactions, epinephrine toxicity presents the clinical picture of tachycardia, arrhythmia, hypertension, chest pain, headache, nausea, and peripheral vasoconstriction. However, as described above, we have used very diluted anesthesia (0.1%), so this toxic reaction is avoided easily.

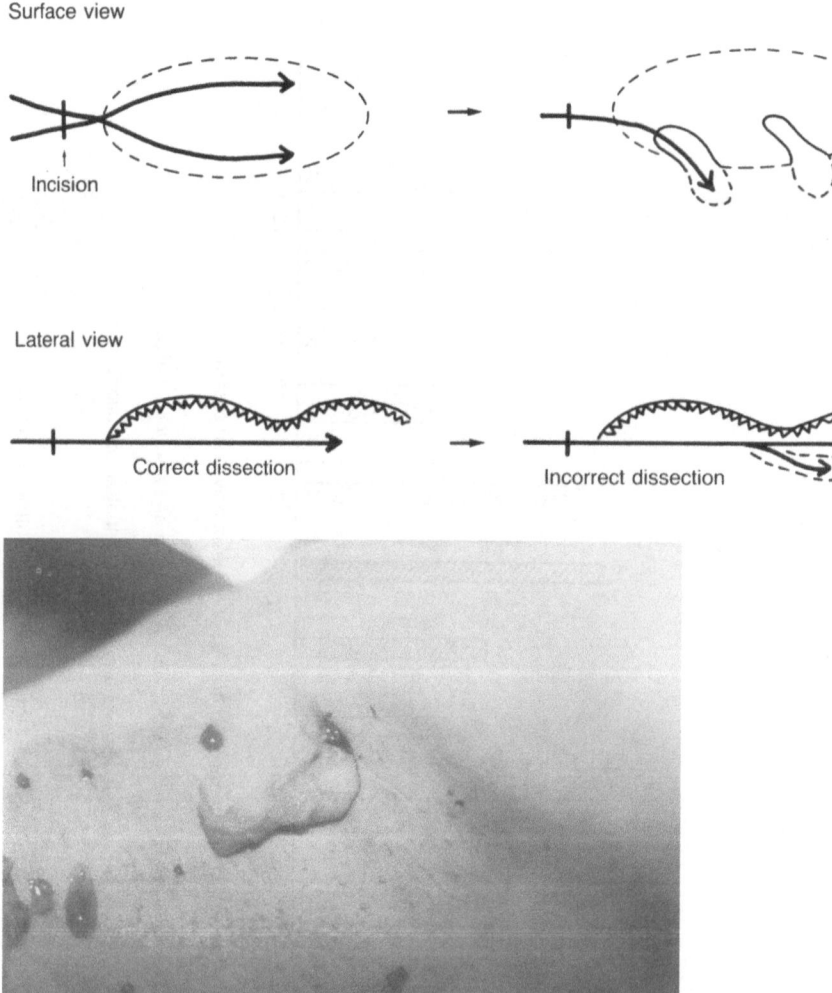

Fig. 15.18.a,b. Dissection by scissoring must be performed inside the marked borderline and must be as thin as possible on the lateral side. **a** Direction of the dissection. **b** Polyp as a result of incorrect dissection. If ignored, hematoma will occur

15.9.2 Dissection and Shaving

Dissection by scissoring must be performed inside the premarked borderline and must be as thin as possible on the lateral side. If not, polyp-like tissue may remain as a source of interference with pressure and fixation, causing hematoma and necrosis (Figs. 15.18.a,b).

If a swelling (bulge) exists in the middle portion of the axillary region, especially in obese patients, a dissection is accomplished under the fibrous fatty tissue so that the skin graft obtains satisfactory adhesion and does not peel off during motion. This deep dissection is ordinarily no problem, since there is little bleeding, but dissection into two layers must be avoided (Fig. 15.18.a).

Proper use of the shaving instrument involves speed, correct pressure, the right grade of blade, and the degree of skin tension. Pressure is gradually added until the subcutaneous tissue is fully shaved. Skin tension is important. If necessary, this tension is obtained by pressing against the chest. Skin damage results in improper tension and injury inflicted by the shaver blade.

15.9.3 Hematoma

If the double tie-over dressing is inexact, a hematoma sometimes occurs, causing necrosis and ulceration. Suitability and unsuitability of subcutaneous scoop suture are very important factors in the double tie-over dressing to avoid hematoma formation.

Causative Factors. Since this method applies an incision of only 1 cm in size and no hemostasis is done to the operated area, the double tie-over dressing is indispensable. Therefore, in order for the recipient area to receive sufficient pressure through dressing A by the tie-over, the subcutaneous scoop suture must be deep and avoid the vascular and nervous systems (Fig. 15.19.a).

Even if the subcutaneous scoop suture is deep and the ligature is tightly contracted, the skin grafts become stuck to undertissue and the lateral side of the graft becomes inflated (Figs. 15.19.a,b). If the subcutaneous scoop suture is shallow and the pressure of dressing A is insufficient, hemostasis becomes incomplete. If the subcutaneous scoop suture in such a condition is tightly contracted, the lateral sides of skin grafts become wrinkled (Figs. 15.19.c,d).

Pressure of Dressing A. If the width of the subcutaneous scoop suture is too wide, the pressure of dressing A is

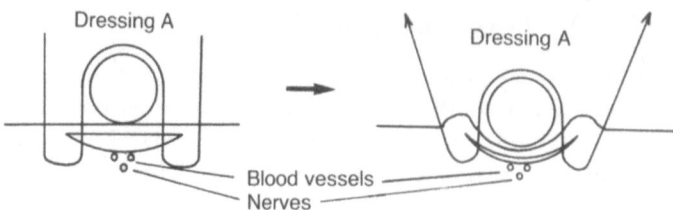

Dressing A   Dressing A

Blood vessels
Nerves

The subcutaneous scoop suture is deep and the pressured supports of dressing A are in an ideal
a condition

The shaved skin becomes stuck to undertissue and lateral sides become inflated. Dressing A is firmly pressed downward, compressing the blood vessels

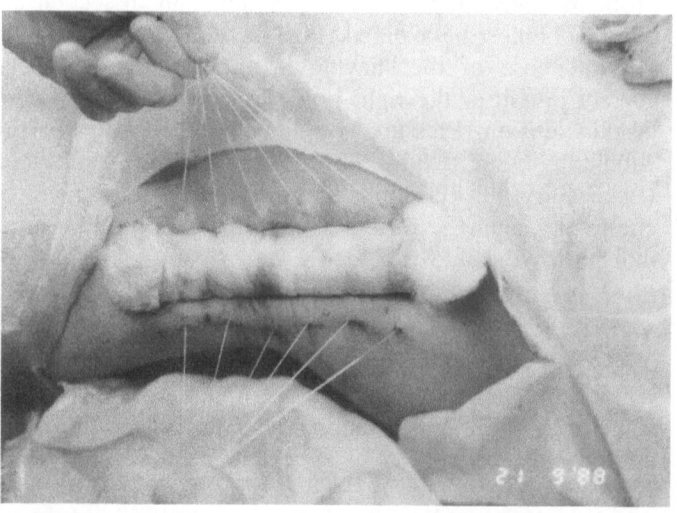

b

Fig. 15.19.a–d. Suitability and unsuitability of subcutaneous scoop suture. **a** Correct method. **b** Correct method with deep suturing.

insufficient and exfoliation of the shaved skin is likely to occur because of the gap (Fig. 15.20.a). On the other hand, if the width of the subcutaneous scoop suture is narrower than that of dressing A, the pressure of the dressing becomes either too strong or too weak (Fig. 15.20.b).

Occurrence. When the dressing is contaminated with blood, especially subcutaneous bleeding in the surrounding healthy skin, redressing must be done as soon as possible. Redressing is minimal, however, due to the basic excellence of this dressing (Fig. 15.21.a). If only one side of the dressing is contaminated with blood, with no subcutaneous bleeding,

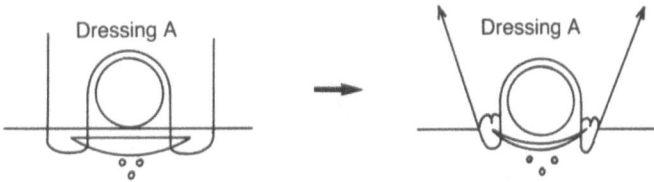

Dressing A → Dressing A

The subcutaneous scoop suture is shallow and the pressure of dressing A is insufficient, making hemostasis incomplete

The lateral sides of the shaved skin become wrinkled

c

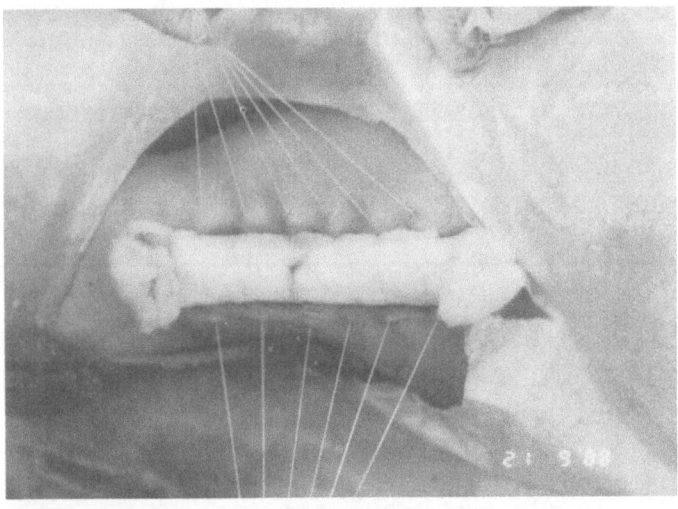

d

Fig. 15.19. (*contd.*) **c** Incorrect method. **d** Incorrect method with shallow suturing

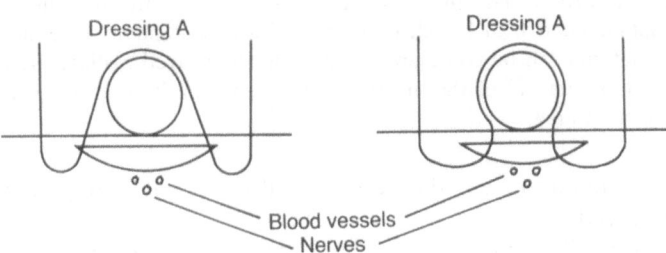

Dressing A   Dressing A

Blood vessels
Nerves

A wide subcutaneous scoop means there is insufficient pressure and
a   exfoliation may occur

A narrow subcutaneous scoop means that the pressure is either too strong or too weak   b

Fig. 15.20.a,b. The pressure of dressing A. (Reproduced with permission from Inaba et al. 1988d)

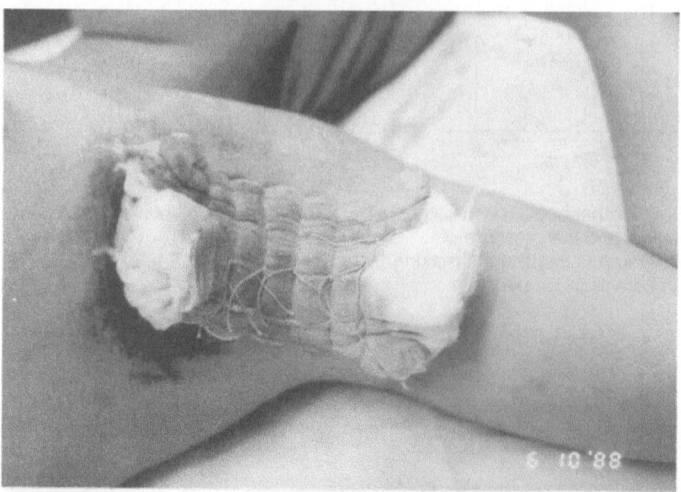

a

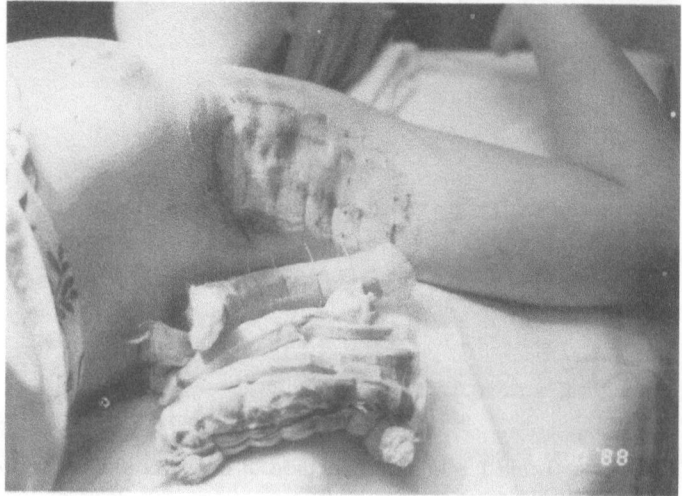

b

Fig. 15.21.a,b. The presence of hematoma. **a** If the dressing is contaminated with blood, especially with subcutaneous bleeding, hematoma can be considered to exist in the shaved axillary area. **b** After removal of the dressing, hematoma can be seen close to the chest area

we continue to use the dressing until it can be completely removed.

A small hematoma (less than 3 mm) discovered when the dressing is removed presents no problem because of the so-called bridging phenomenon, and necrosis does not occur (Fig. 15.21.b). If a solitary hematoma of more than 1 cm occurs, the affected part is excised using a hematome to exclude the blood coagulum and is suppressed slightly with gauze.

Should a large hematoma or shaved skin cyanosis be observed in graft skin, redressing must be immediate. If the hematoma is dealt with, the shaved skin can still survive, even if the hematoma is observed when the dressing is completely removed (after 6 days). Redressing is recommended, with the hematoma completely excluded.

Fibrinolytic therapy on hematomas beneath grafted skin is also very efficacious. When a hematoma is observed beneath the grafted skin, streptokinase (Varidase, Lederle Co., Tokyo) is used in a concentration of 5000 U/ml under a pressure bandage for 6 h. This method has proven to be valuable in facilitating the removal of a hematoma beneath grafted skin and in improving the results of skin grafting (Ono et al. 1978).

Huge Hematoma After Removal of Dressing. In some cases a huge hematoma occurs immediately after the dressing is removed (Figs. 15.22.a–c). This hematoma results from the separation of the polyp-like tissue from the lateral side or a lack of smooth dissection complicated by arm motion in spite of bleeding having been completely suppressed by the dressing pressure. In this case, the central axillary sector must be incised along the wrinkling. After the removal of a sizeable hematoma, the presence of polyp-like tissue or a lack of smooth dissection in the recipient region must be investigated. Redressing is done after the removal of this protruding tissue or suturing of the deflated region (Fig. 15.23).

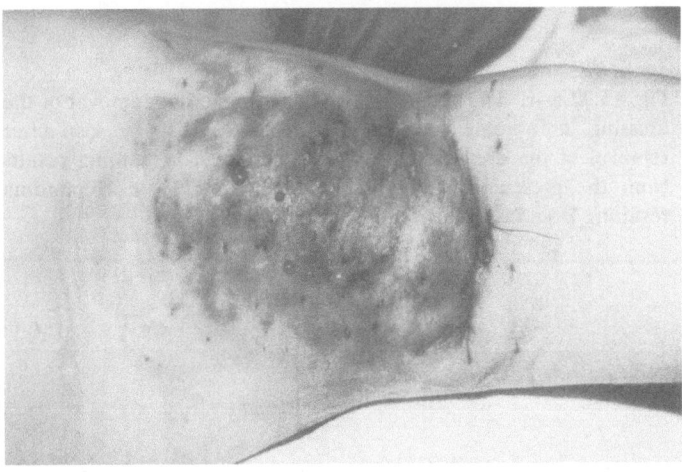

Fig. 15.22.a–c. The occurrence of hematoma after removal of the dressing

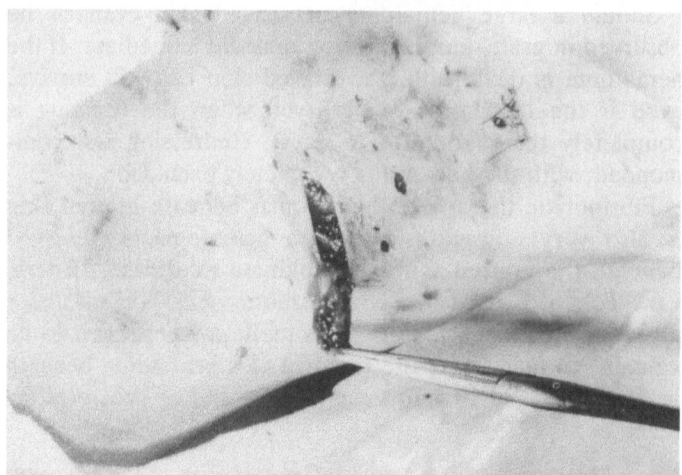

b

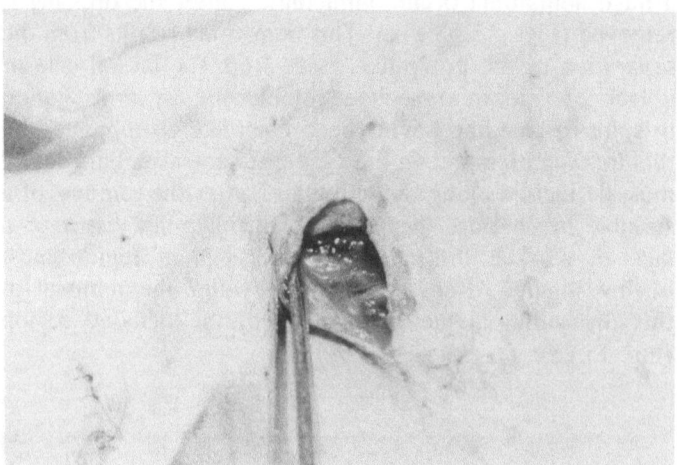

c

Fig. 15.22.a–c. The occurrence of hematoma after removal of the dressing. **a** In some cases a huge hematoma can be seen after removal of the dressing (case 1). **b** This huge hematoma results from the presence of polyp-like tissue (case 1). **c** Hematoma resulting from lack of smooth dissection (case 2)

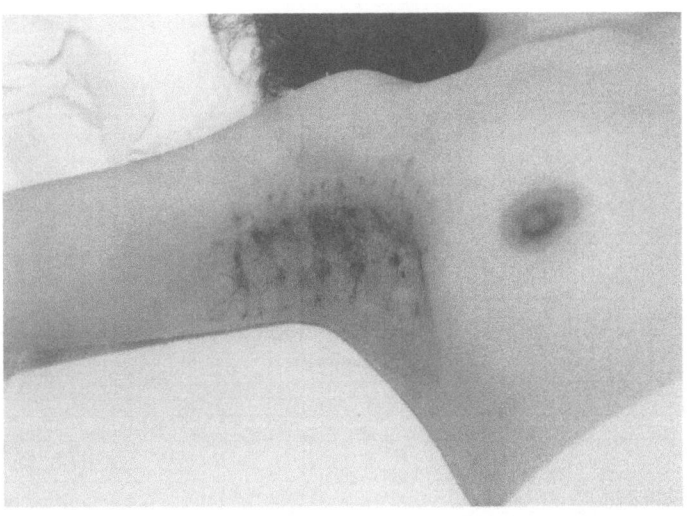

Fig. 15.23. Redressing must be done after the removal of polyp-like protruding tissue or suturing of the deflated region. Shaved skin is grafted completely without necrosis

15.9.4 Infection

The structure of the double tie-over dressing acts to control pressure. Following the operation, a hardened fixation (dressing) is needed for hemostasis. On the next day, three layers of the central gauze portion from the top of dressing B must be removed to reduce the pressure. If the suturing thread turns back inside the skin, the grafted skin tends to become infected, resulting in ulceration. However, use of antibiotics, including chloromphenicol ointment, when employing the double tie-over dressing shows good anti-bacterial results. Before the use of antibiotics, inflammation occurred at a frequency of 1%–2%, but with this procedure there is almost no inflammation and ulceration.

15.9.5 Ulceration

If ulceration does occur due to inflammation, hematoma, or over-pressure, it will tend to widen in spite of using even stronger antibiotics. Patch grafting as soon as possible shows good results (Figs. 15.24.a–d). This patch grafting (postage stamp graft) is the same as a thin split-thickness graft subdivided into small "postage stamp" squares and applied to the ulcer. The graft rapidly forms epithelialization to prevent a recurrent ulcer. This type of graft can be

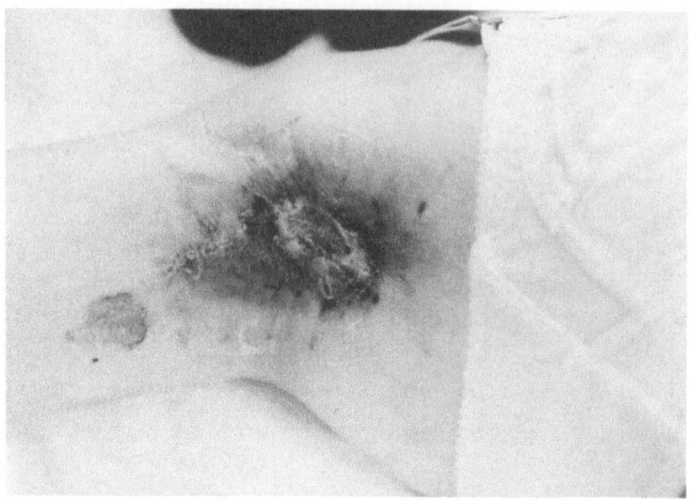

a

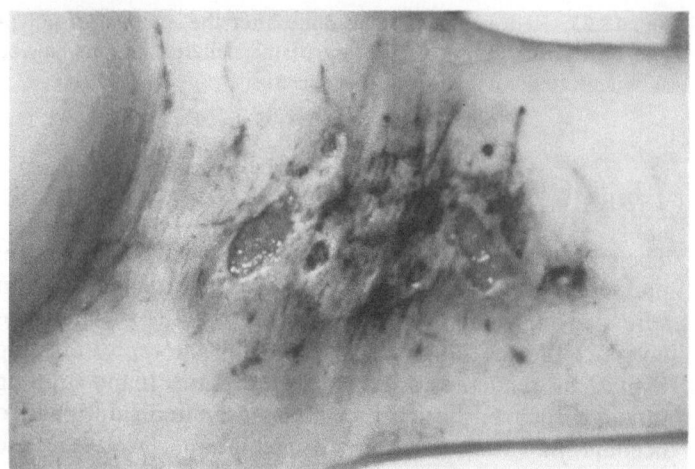

b

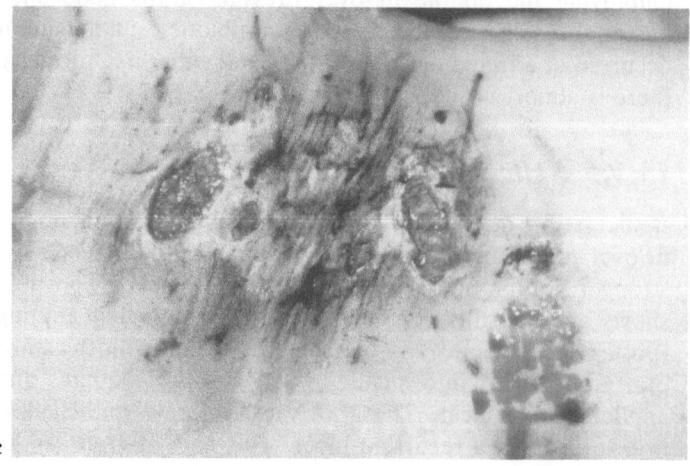

c

Fig. 15.24.

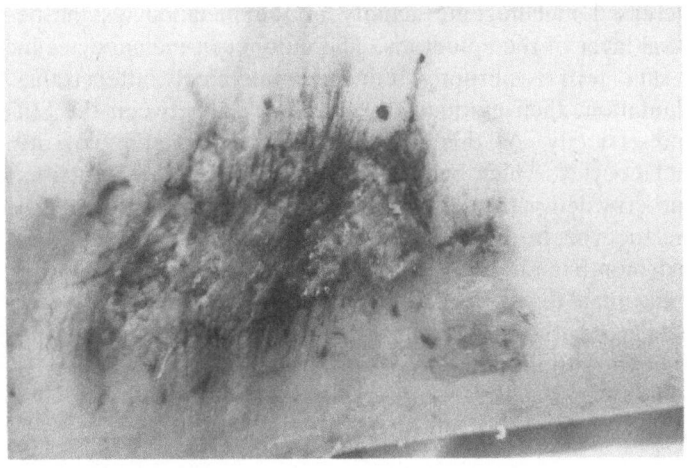

Fig. 15.24.a–d. Patch grafting. **a** Patch grafting shows good results. The graft rapidly forms epithelialization only 3 days after grafting. This patch graft is taken from the arm side. **b** Ulceration due to inflammation. **c** Patch graft used on ulceration even as inflammation persists. **d** Rapid improvement is evident in epithelialization

useful even if considerable drainage or exudation in the wound is necessary, and inflammation may be counteracted. If the ulcer is very large, patch grafting is useful to cover the ulcer, but tends to contract by defection of the dermis. If a deep ulcer is formed, it can be restored with patch grafting up to the level of the normal epidermis, and is flattened to the normal level in an average of 20 days. This finding indicates that a thin-thickness graft (patch graft) has an epidermal stimulatory factor.

15.9.6   Hyperpigmentation

Hyperpigmentation often occurs more frequently among nonwhites in skin grafts after transplantation. It is accepted that melanin pigmentation reflects the melanin content of the keratinocytes. Apparently melanin pigmentation is related to the following biological processes: after the production of melanosomes, they are then transferred from melanocytes to keratinocytes, leading to the degradation of the melanosomes. All of these processes appear to be influenced by genetic factors.

Various exogenous stimuli, e.g., X rays, ultraviolet rays, and the stripping of skin, produce hyperpigmentation by increased melanogenic activity of the melanocytes in the basal layer of the epidermis. The number of melanocytes in a skin graft is abruptly reduced immediately after transplantation, then gradually reaches a peak between the 8th and 21st day. At this stage, there is hyperactivity of the melanocytes, which become larger in size, highly dentritic, and crowded with melanin granules. The mean numbers of melanocytes in the graft undergo a gradual decrease in the 3rd month and, for the next 3 years, remain at a lower value than that before transplantation. In our experience, the pigmentation of shaved skin returns to normal in patients without a keloid constitution.

15.9.7   Keloids and Hypertrophy

Normally, the biochemical processes of cosmetic wound repair culminate in fine scars as the only evidence of dermal injury. In certain individuals, however, the repair progress may go awry and wounds may heal with large, raised collagenous scars known as keloids or hypertrophic scars.

Occurrence. Hypertrophic scars appear to be a self-limiting type of overhealing following injury. With time, the raised, hypertrophic scar will become flat and pale. The mature keloid, however, will extend beyond the confines of the original wound and will not regress. The aesthetic complications of abnormal scar formation are often severe and clinical management is frustrating.

Keloids represent a fibroblastic response to injury in excess of that appropriate for repair of the injury. In many instances, however, a preceding injury cannot be identified. The tendency to develop keloids depends upon a number of factors, including race, location on the body, and type of injury. Blacks and other deeply pigmented individuals generally are more susceptible to keloids than those with fair skin. Moreover, the tendency toward keloids may run in families. The upper chest (especially the sternal area), the ears, chin, shoulders, neck, and lower legs are more susceptible than other areas of the body, and increased natural skin tension appears to play a role. Keloids have a greater tendency to occur following electrodesiccation or cautery procedures than after scalpel surgery. Infection and inflammatory changes render a wound more susceptible to keloid development; for example, the inflammatory reaction in an acne papule on the chest is sufficient to initiate keloid formation. Extensive keloids often follow thermal burns.

A keloid usually begins as a small, firm, erythematous papule which slowly enlarges. Enlargement may be regular, producing a round to oval lesion, or it may be irregular and eccentric with clawlike extension. Further development may lead to large nodular, lobular, or pedunculated lesion. The area of involvement may suffer considerable distortion as a result of extensive keloid formation. The progression from wound healing to keloid formation may become evident several weeks following an injury, and the process may continue for months to years before spontaneous arrest occurs (Figs. 15.25.a,b).

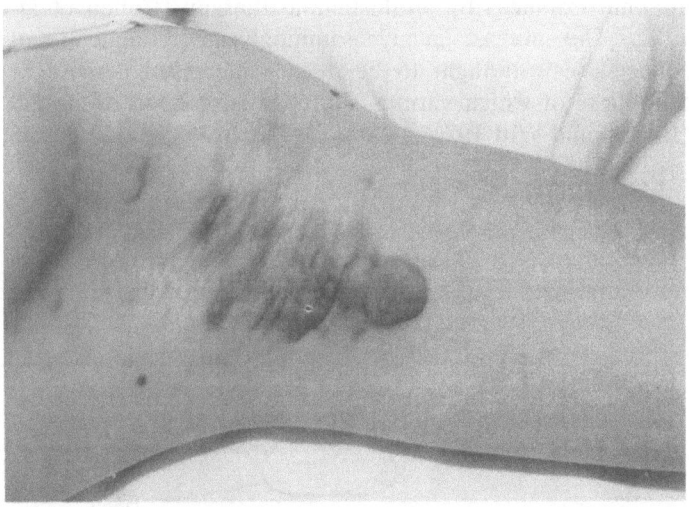

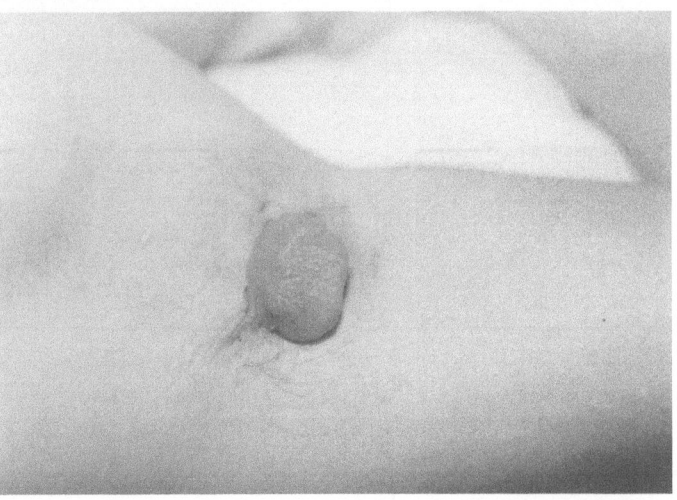

Fig. 15.25.a,b. Postoperative keloid formation. **a** Keloids observed after subcutaneous tissue shaving. **b** Further development of the keloids to large nodular, lobular, or pedunculated tumors

Character. Early keloids have a rubbery consistency and are usually symptomatic, with significant itching, tenderness, or pain. Quiescent lesions are quite firm, asymptomatic, and hyperpigmented. In most pathogenetic studies, keloids show the same biochemical and pathologic abnormalities as in hypertrophic scars. Collagen synthesis is markedly increased and active collagenase has been identified in above-normal amounts (Milson and Craig 1973). Alpha globulin may inactivate collagenase (Cohen et al. 1975). Increased numbers of mast cells may stimulate collagen synthesis, and pruritus in keloids can be abolished in some instances by antihistamine therapy (Cohen et al. 1972). The increase in glycosaminoglycans, collagen, and collagenase is thought to be due to increased activity of fibroblasts of which various subtypes have been described (Russell and Witt 1976).

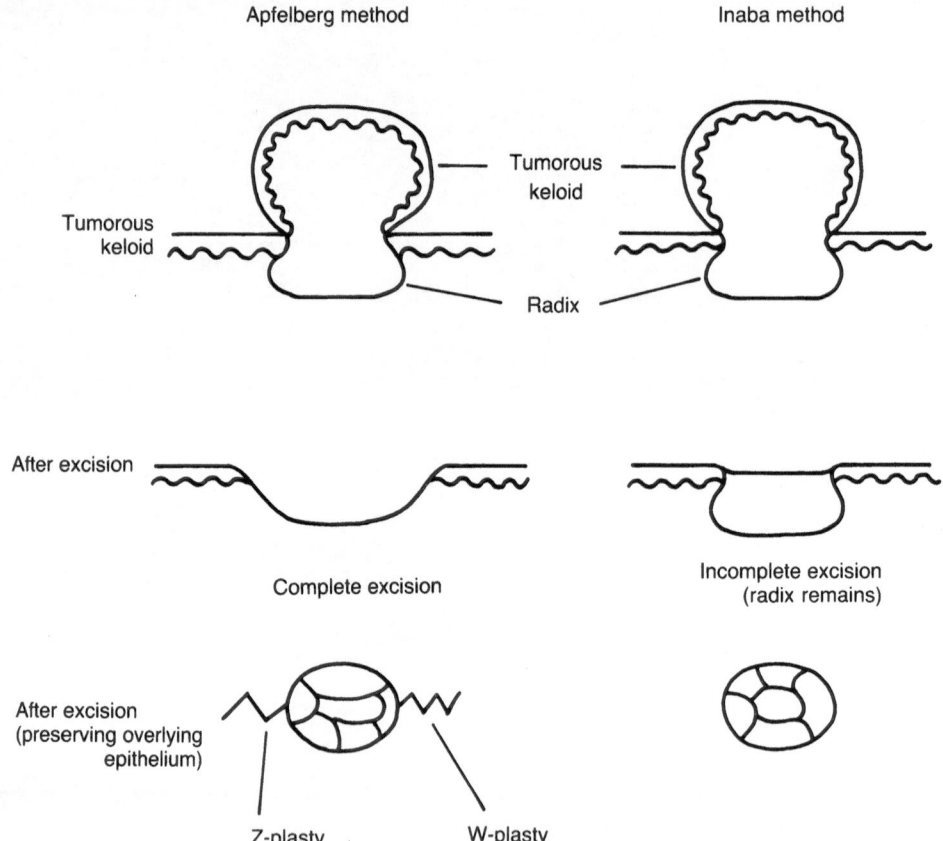

Fig. 15.26. Comparison of the Apfelberg and Inaba methods for treatment of tumorous keloids

Treatment. The most important aspect in the treatment of keloids is prevention. In patients known to form keloids, elective surgical procedures should be avoided, but when necessary, scalpel surgery with careful attention to aseptic technique and avoiding wound tension is mandatory. Electrosurgical techniques should not be used. Simple excision of a keloid often will result in regrowth, but this recurrence may be inhibited or prevented by the use of superficial X-ray therapy to the area following excision (Levy et al. 1976). Keloids also can be reduced in size by periodic intralesional injection of insoluble corticosteroid suspensions, such as triamcinolone acetonide.

We developed a new technique for tumor-type keloid removal which does not result in excessive, unsightly scarring (Inaba et al. 1985b). In other surgical excision techniques (Apfelberg et al. 1976) the entire keloid, including the base (radix), is removed (Fig. 15.26). In our technique, however, the keloid is removed at the level of the dermis leaving the basal part of the keloid to minimize the scar size. The epithelial layer on the surface of the tumor-type keloid is grafted to the recipient site. A local corticosteroid injection is then administered. This technique is fully effective and leaves a clean surface postoperatively (Figs. 15.27.a–d).

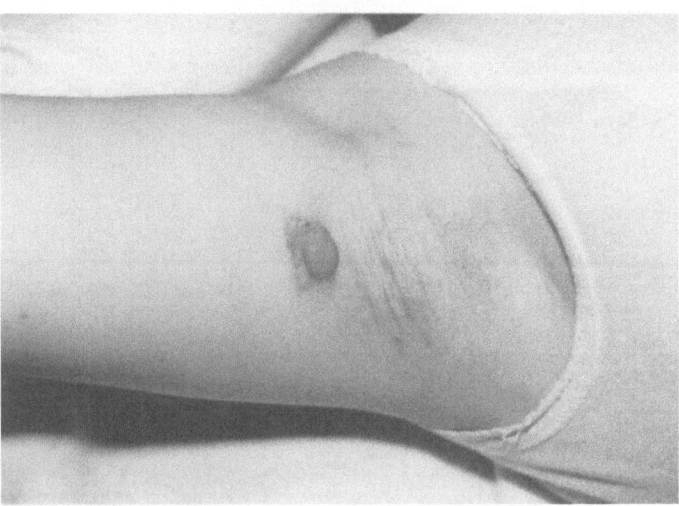

a

Fig. 15.27.a. Keloid removal by the Inaba technique

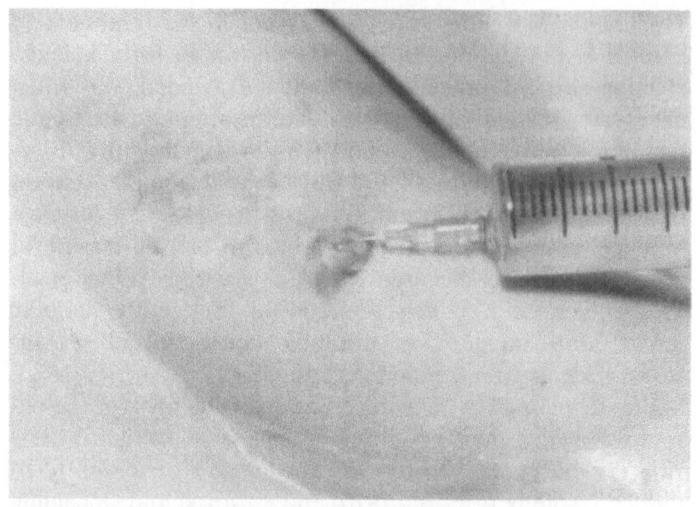

b

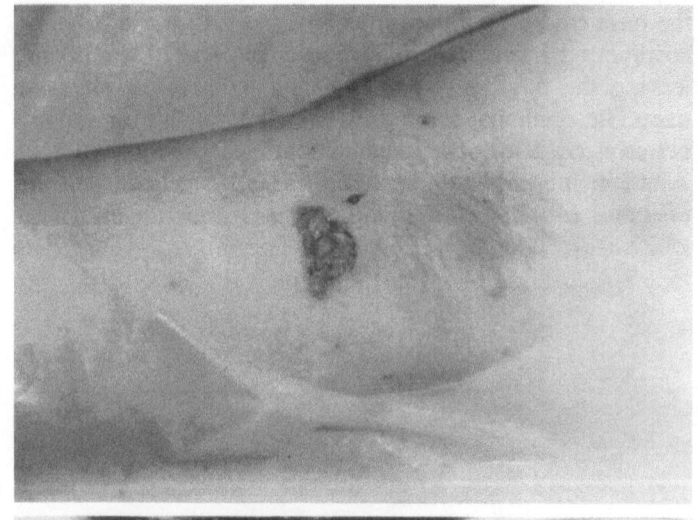

c

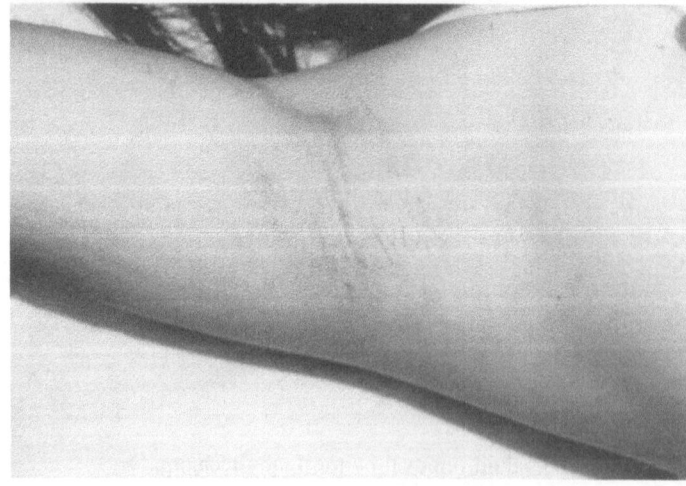

d

15.9.8   Paralysis of Upper Extremities

In the use of the double tie-over method to suppress bleed-ing, the suturing runs from the edge of healthy skin, into the tissue below the skin graft, to the edge of the dressing A and over the applied dressing, continues to the other side of the normal skin, and is tied up over the dressing (Fig. 15.10). If this suturing is too shallow, it cannot suppress the blood vessels. It must go in deeply; the support points are very important. The needle may occasionally injure the brachial nervous plexus and blood vessels.

We have encountered 7 cases of nerve dysfunctions. In one case, the application of the double tie-over dressing brought severe pain, but the patient endured it. Imme-diately after the operation, paralysis was observed in the middle and little fingers, and the dressing was replaced with another type. About 12 months were required for complete recovery (Fig. 15.28). Radial paralysis was observed in three cases (thumb and index finger paralysis and par-esthesia), medial paralysis in two cases (paresthesia), and ulnar paralysis in two cases (paresthesia and no unfolding of the little finger) among the 20,000 cases treated to date. Damage, however, is limited to a nerve invaded by a needle; within 5–8 months the disorder disappears, so it is not a long-term problem.

In some cases, when radial paralysis was observed in either the right or the left hand of a patient immediately after the operation, they were given a dosage of methycobal and advised to massage the hand well while taking a bath during the next 6 months for a complete cure. One patient visited the clinic after 6 months and explained that the paralysis remained and he was unable to work. He had consulted a specialist and had been told that the only way to cure it was to undergo a neurolysis operation, but since it was rather late to apply that treatment, he had not been assured of a successful result. The patient had become upset and visited our clinic with a demand for compensation. We attempted to ease the patient's distress by explaining that

Fig. 15.27.a–d. Keloid removal by the Inaba technique. a Typical keloid formation at the incision site. b Superficial anesthetic injec-tion administered to obtain thin-thickness graft. c The keloid is removed at the level of the dermis, with the basal portion of the keloid remaining. Thin-thickness graft obtained from the surface of tumor-type keloid is grafted. d Appearance of the axilla two years postoperatively, following corticosteroid injections

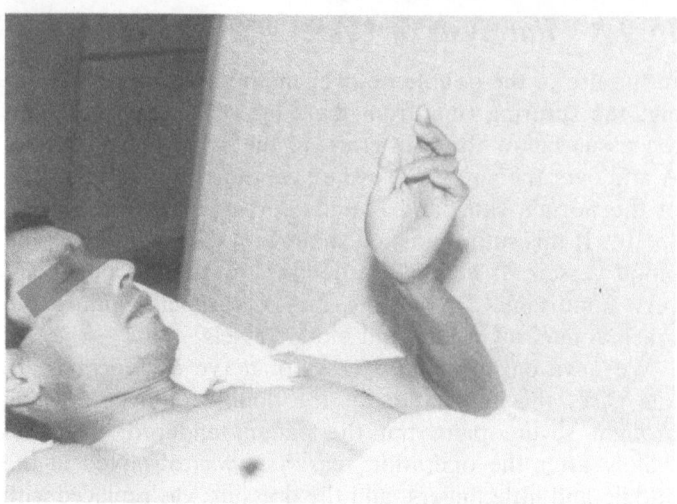

Fig. 15.28. Ulnar paralysis, medial nervous disorder. After one year, movement is normal, but paresthesia persists

the paralysis caused by ligation is only temporary and asked him to observe the condition for another 3 months. We insisted that if the patient underwent an operation performed by any other physician, we would take no responsibility for the result. After 3 months, the patient returned with fairly free movement of the right-hand fingers and did not need to come back for further consultation.

Paresthesia can sometimes be seen superficially at the upper arm region after operation but this is not a problem.

Patients receive anesthesia in both axillae. At this time the arms sometimes become paralyzed, and the patients cannot maneuver their hands above their heads. This causes overextension of the arm, resulting in over-suppression of the brachial nerves by the pectoralis minor (one of four muscles of the anterior upper portion of the chest). It may be correct at this time to release the over-extension (Figs. 15.29.a,b).

15.9.9 Formation of Cysts, Acne Vulgaris, and Atheromas

Cysts, *acne vulgaris*, and atheromas are sometimes formed on the shaved skin postoperatively. The causative mechanism is not clear, but these disorders may be due to the grafting condition, because they almost always occur in the induration area as a result of incorrect grafting. Studies of

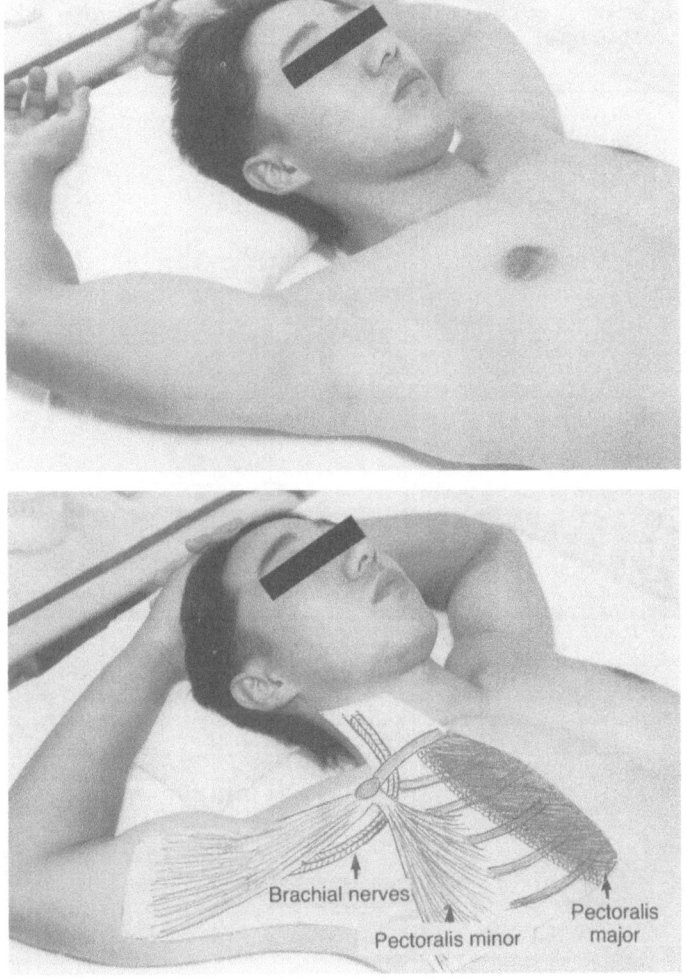

Fig. 15.29.a,b. Releasing the postoperative overextension of the arm. **a** The arms sometimes become paralyzed and cannot be maneuvered above the head. **b** Overextension of the arm results in over-suppression of the brachial nerves by the pectoralis minor

the causative mechanism (Leyden and Shalita 1986) point to: (1) hyperkeratinization and obstruction of sebaceous follicles, resulting from abnormal desquamation of follicular epithelium, (2) an androgen-stimulated increase in the production of sebum, (3) proliferation of *Propionibacterium acne*, which generates inflammation, and (4) disruption of the preclinical precursor lesion (the microcomedo), producing inflammation which leads to the pustules and

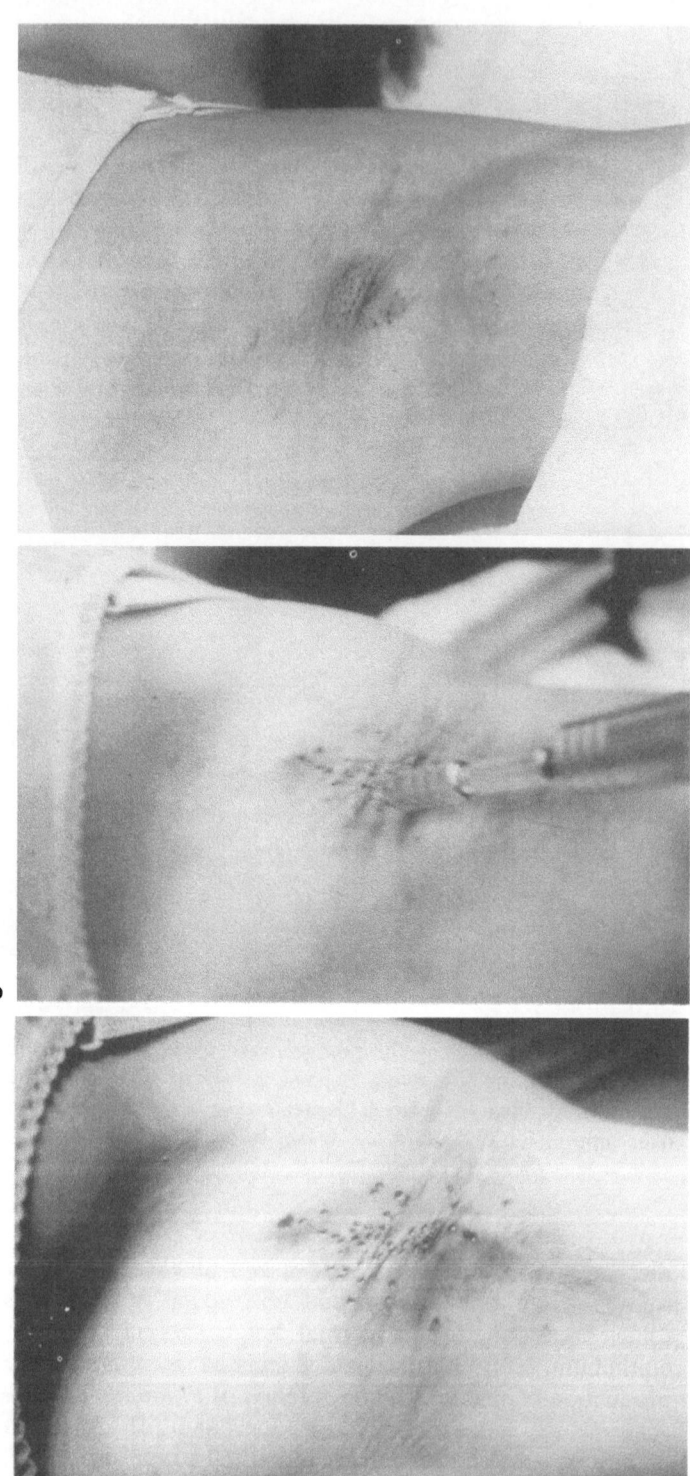

Fig. 15.30.

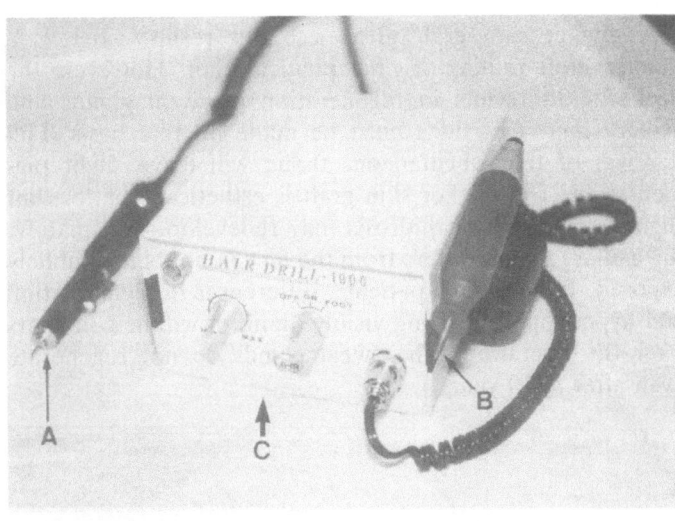

Fig. 15.30.a–d. Good results from the small-diameter punch method in treating postoperative acne-like disorders. **a** Small epithelial tumor following surgery. **b** The small-diameter (1 mm) punch method is used in order to remove the appendage of the hair follicle completely. **c** After removal of the appendage of the hair follicle. **d** Instruments. *A*, 1-mm diameter micropunch; *B*, punch for hair insertion; *C*, hair drill 1000 controller

papules of clinical disease and may eventually result in scarring.

To treat the acne-like disorder, we have used a small-diameter (1 mm) punch method to remove the appendage of the hair follicle completely. Good results have been obtained without scarring (Figs. 15.30.a–d). Further research into the causative mechanisms seems to be warranted.

15.9.10 Follow-up Results

We reported as early as 1974 (Inaba et al. 1974) on follow-up results of the subcutaneous tissue-shaving procedure. Results had been very efficacious in 92.6% (187 cases) of the 202 patients, acceptability efficacious in 4.5% (9 cases), and not efficacious in 2% (4 cases) of the patients (see Table 14.2). This operation leaves only a 1-cm scar from the small incision made to insert the instrument as well as several other minute incisions (0.2–0.3 mm) made to evacuate blood. The remaining scars are not a problem for patients with a nonkeloid constitution. After 6 months to 1 year, the scars are no longer discernable.

A thicker skin graft gives a better result, whereas a thinner graft retains original pigmentation. However, the thicker graft results in regeneration of sweat glands and axillary hair. Decisions must be made on that basis. Full removal of the subcutaneous tissue will leave slight pigmentation. The thicker skin graft is esthetically better, but slight odor and hyperhidrosis may redevelop. Fortunately, the axillary region differs from the face in not being publicly exposed. In almost all patients, differences in pigmentation and hypertrophic scarring visibly diminish within 1–2 years after the operation. The sweat glands do not regenerate even after 5–10 years.

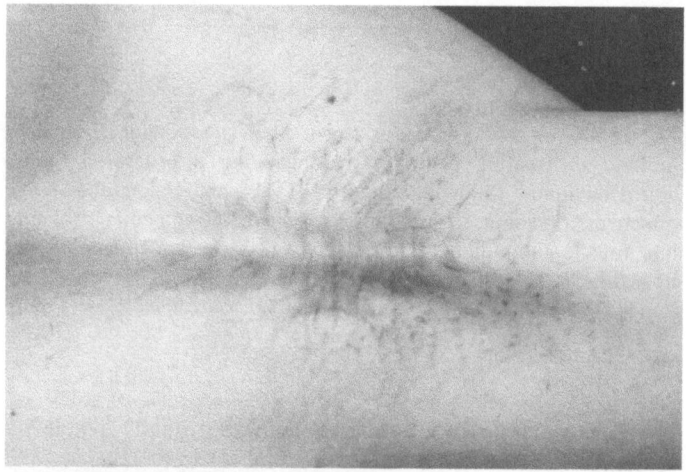

a

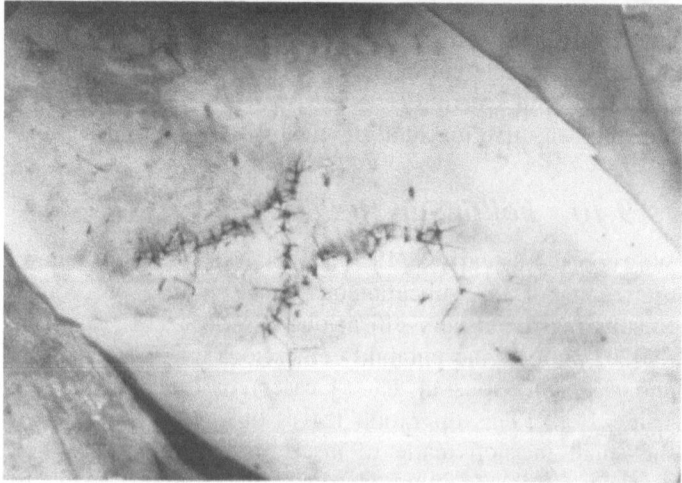

b

Fig. 15.31.a,b. Postoperative scar constriction. **a** Constriction of the axilla caused by incomplete excision. **b** Z-plasty surgery to correct constriction

15.9.11   Treatment of Scar Constriction Cases

There are some cases in which constriction is observed after performing the incomplete removal method (Fig. 15.31.a). In such cases, a Z-shaped surface incision is made over the skin in order to turn it over and remove the sweat glands and the hair roots by using the clearance method (Inaba et al. 1988c). This is the Z-plasty formation method (Fig. 15.31.b).

15.9.12   Upper Arm Dysfunction

Patients show concern about holding the upper arm properly and about not being able to lower the upper arm to the chest because of the pressure dressing. They are also concerned about not being able to work for 1 week. However, the patient can actually resume work 3 or 4 days after the operation when the dressing has been decompressed. Some patients claim to experience difficulty in holding up the upper arm. This phenomenon can occur if the patient does not elevate the upper arm after the operation, so that the skin adheres to the lower muscle in a contractive manner. We recommend elevating the arm as vigorously as possible 10 days after the operation.

15.9.13   Treatment After Other Incomplete Procedures

Many patients treated with incomplete procedures later visit our clinic for the subcutaneous tissue-shaving operation (Inaba et al. 1988c). In those cases in which large scars are present, the cutting edge of the blade beneath the scar tissue is too sharp to be effective on extensive hard tissue caused by incomplete treatments such as the removal method. It is efficacious in cases with relatively less scarring left, for example, after the electrocoagulation or curettage methods.

We attempt to reconstitute the scarred epidermis after removal of the scar tissue, and remove the remaining sweat glands in surrounding hair-bearing scar tissue. For this reason, it is necessary to eliminate as much of the scar tissue as possible. We have devised a pair of special scissors which can easily remove the tissue. The distal edge of the blade used for snipping off the scar tissue is blunted by about 1 mm. Snipping readily separates the underlying scar tissue embedded in the skin and then removes the scar tissue, leaving the epidermal surface intact (Figs. 15.32.a–c).

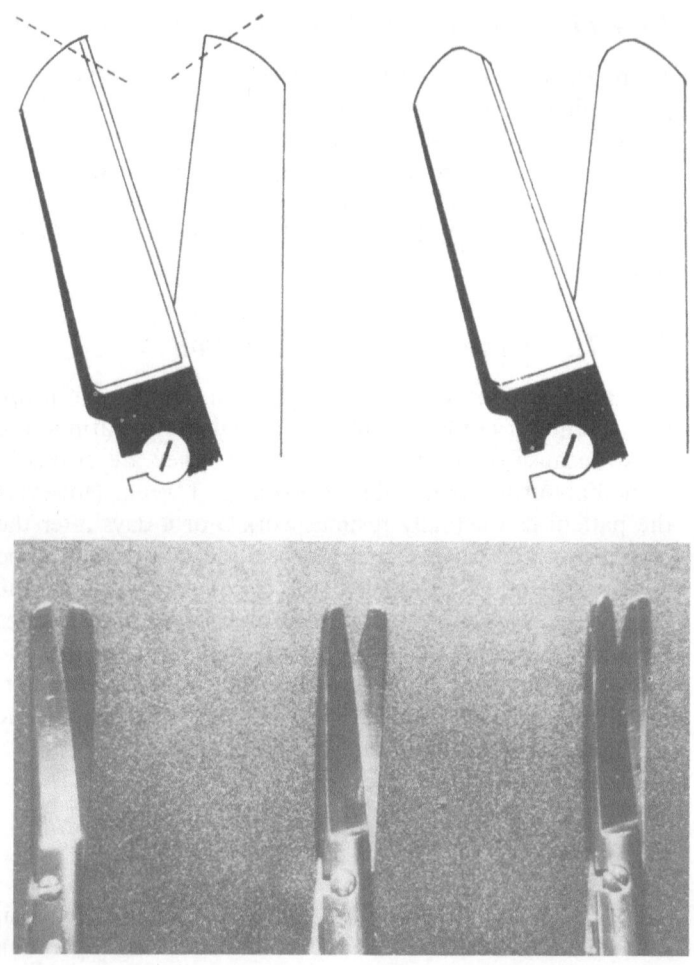

Fig. 15.32.a–c. Treatment of incomplete removal of scars. **a** Development of a new type of scissors to remove scar tissue. The distal edge of the scissors blade is blunted by about 1 mm. This makes it easy to separate and remove scar tissue embedded in the skin. **b** Blunt-tipped scissors used to remove scar tissue. Scissors *left* are not beveled; *center*, partially beveled; *right*, fully beveled. **c** Removal of the scar tissue using blunt-tipped scissors

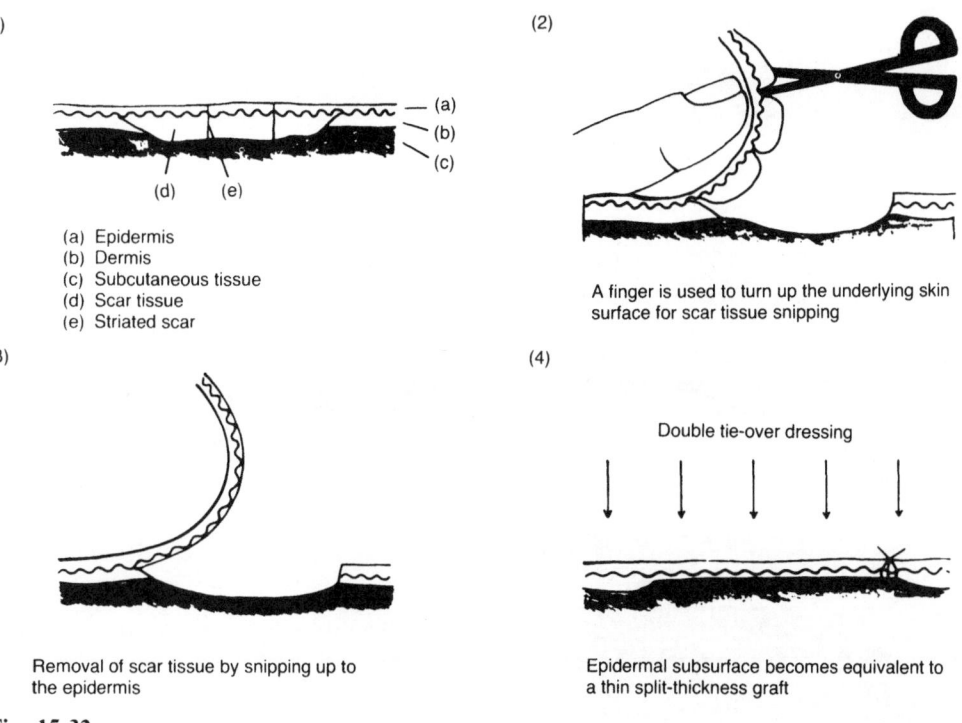

(1)

(a) Epidermis
(b) Dermis
(c) Subcutaneous tissue
(d) Scar tissue
(e) Striated scar

(2)

A finger is used to turn up the underlying skin surface for scar tissue snipping

(3)

Removal of scar tissue by snipping up to the epidermis

(4)

Double tie-over dressing

Epidermal subsurface becomes equivalent to a thin split-thickness graft   c

Fig. 15.32.

By turning back the undersurface with the fingers and carefully snipping the underlying scar tissue from an additional central portion incision along the wrinkling in the axillary region, this approach enhances the appearance of previous scars. After snipping off the scar tissue, we use the subcutaneous shaver to prevent regeneration of sweat glands. Following this procedure, the texture of the scar surface becomes much like that of a healthy skin surface. The removed scar tissue leaves a subsurface area similar to a skin graft. This removed skin is then grafted in place by the double tie-over method.

Excellent results in radical treatment of incomplete procedures have been obtained with this special procedure. It does not, as such, suffice to remove wrinkling which has radiated from the edges of the scar. To deal with these wrinkles, an incision is used to spread each wrinkle apart. A patch graft can then be used to conceal these wrinkles and further improve the appearance of the skin surface (Figs. 15.33.a–c, 15.34.a–c). We believe that these procedures for treatment of scarring and wrinkling can be applied not only to the axilla but to other areas where normal physical motion could be a postoperative problem.

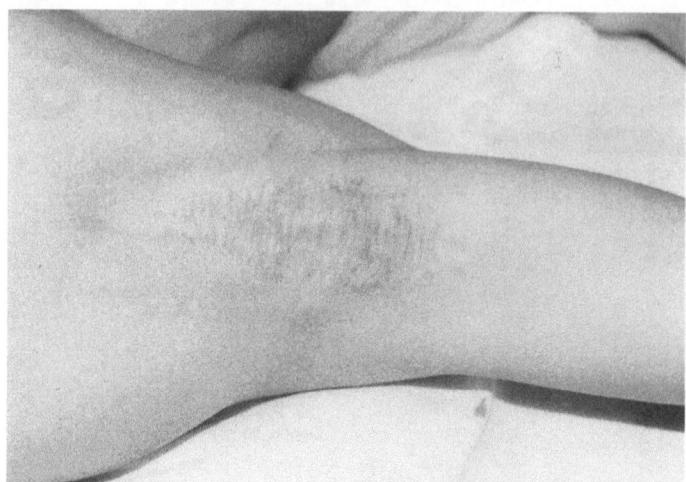

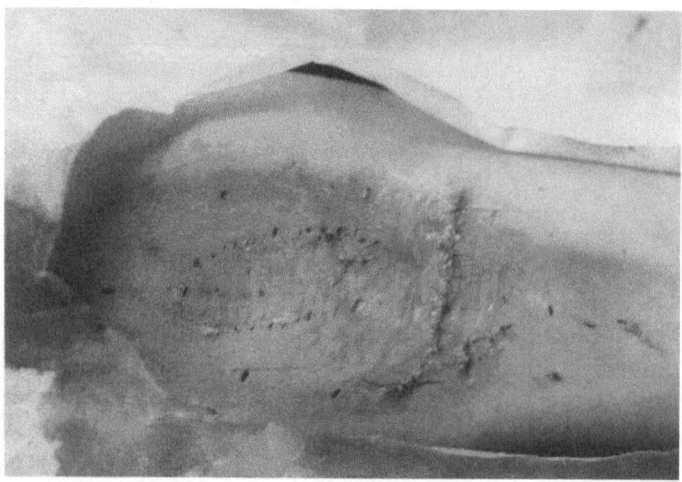

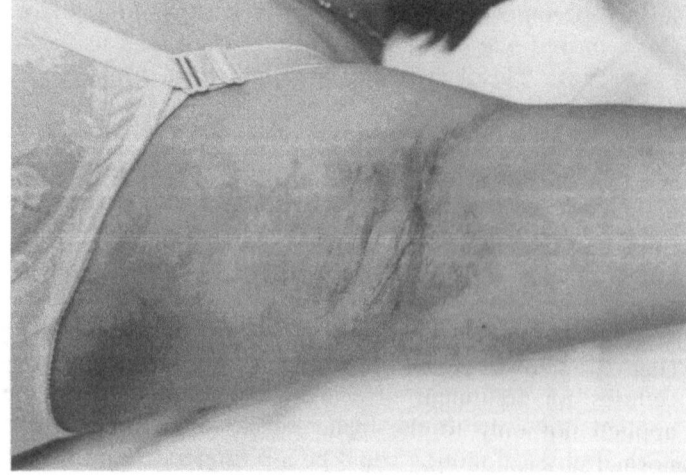

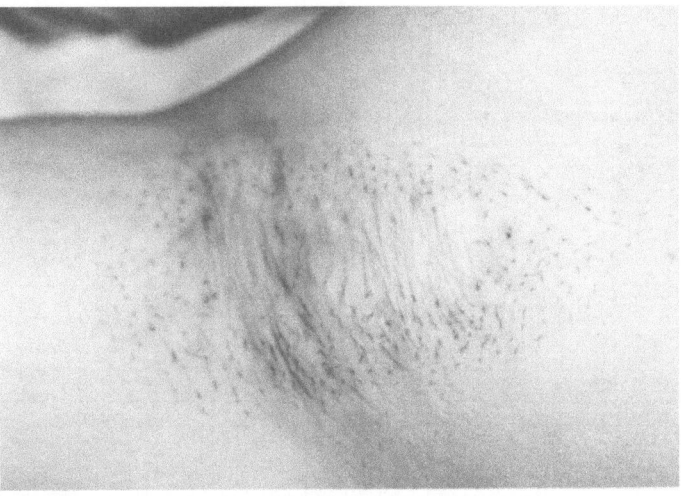

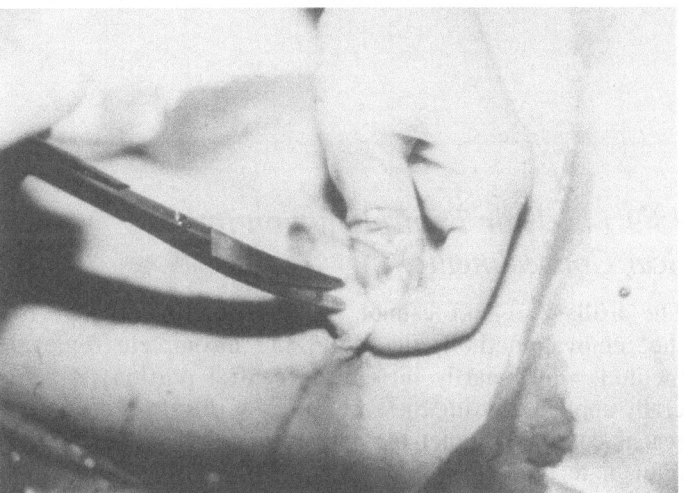

Fig. 15.34.a–c. Procedure for reducing postoperative wrinkling. **a** Incomplete removal method. The scar has widened and much wrinkling is observed. **b** Specific scissor removal of the scar tissue. **c** After this treatment, almost all wrinkles are diminished. Surrounding axillary hairs are removed completely

Fig. 15.33.a–c. Procedure for reducing postoperative wrinkling. **a** Incomplete removal method. The scar has widened and much wrinkling is observed. **b** To deal with these wrinkles, after the removal procedure, a small incision is used to spread each wrinkle out. **c** One year later almost all wrinkles are diminished. This large scar may also diminish over time; remaining odor and hyperhidrosis have diminished

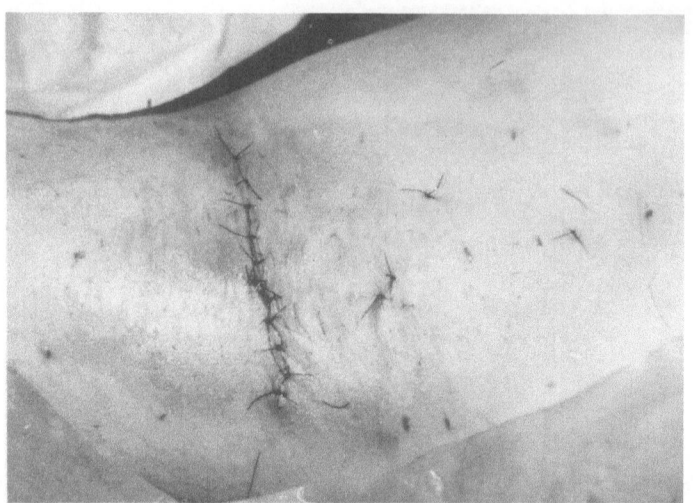

Fig. 15.34.c.

15.9.14   Hair Transplantation for Scar Concealment

The axillary region cannot be completely removed by the removal method; therefore, the incomplete removal method, which partly incises the central portion, is generally applied. During the early stage of this procedure, the scar is concealed under the axillary hair which presents no immediate problem. However, with upper arm movement and influence of muscular strength, the axillary skin will extend and the scar becomes enlarged in a manner similar to the preoperative condition under muscular tension.

Axillary hair remains around the scar, and, as a result, an ugly scar surrounded by many lines is left at the center. There are some patients who wish to undergo hair transplantation in order to conceal the scar (Fig. 15.35.a).

In hyperhidrosis, the skin is shaved to the degree of a medium split-thickness skin graft which is so thin that it prevents hair regeneration. This is ideal for female patients, yet embarrassing for male patients, who then wish to undergo hair transplantation to restore a musculine appearance (Fig. 15.35.b).

We collected scalp hair and pubic hair with a small punch for transplantation but discovered that the punched parts of the underarm would become enlarged, making insertion

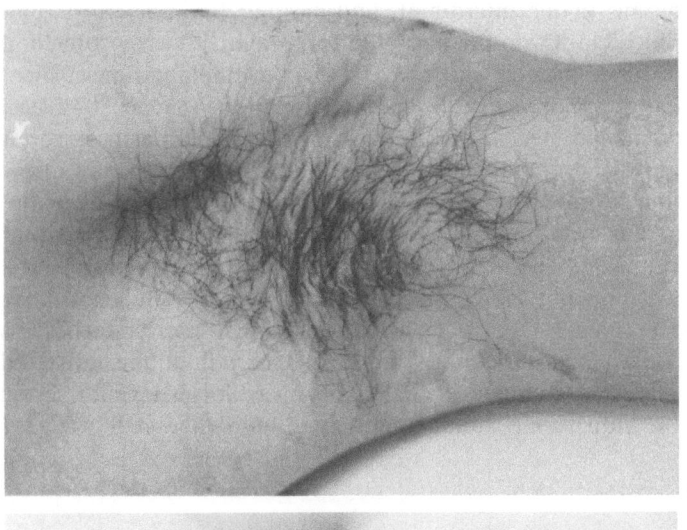

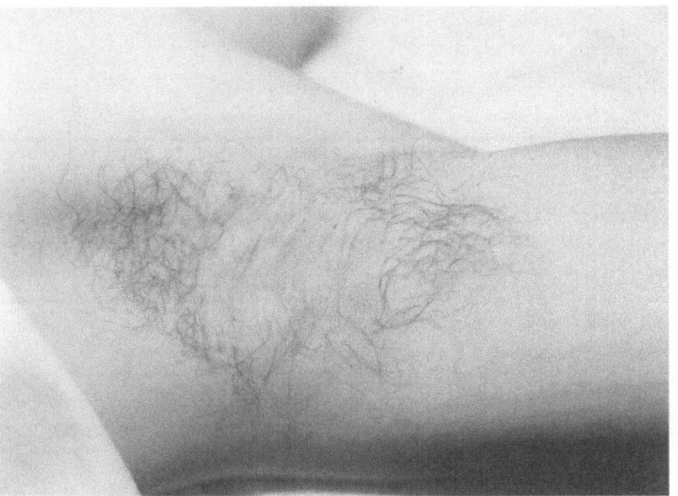

Fig. 15.35.a,b. Hair transplantation to conceal scars. **a** Single-hair transplantation used to conceal the scar. **b** Complete removal method with remaining axillary hair. Male patients wish to undergo hair transplantation to restore a masculine appearance

difficult. We have developed a comparatively easy-to-perform single-hair implantation method to conceal scars of the scalp, eyebrows, and pubic area. Scalp hairs are collected by firstly cutting them to a length of 5 mm and then setting them inside a 1 mm diameter micro punch instrument (MHR Co., Ltd., Tokyo) (Fig. 15.30.d). The essential point here is to punch the scalp parallel with each hair up to the dermal level so that the hair can be collected in perfect shape. Then a pincette is used to hold the epi-

dermis of the micro-graft and another fine pincette is employed to lift up the hair bulb for a relatively easy collection
of single hairs (Figs. 15.36.A–C). For implantation, a specially developed needle for hair insertion (gauge 20–18) is
employed (Figs. 15.36.D,E). This needle is partly covered
in order to prevent the hair from being pulled out at insertion. Prior to the implantation, punches are made over the
area of insertion by the punch instrument for hair insertion.
If this procedure is omitted, hairs over the normal area
will pop out along with insertion. The advantages of this
method are: (1) the micrografting (single hair collection) is
not so painstaking, (2) no scars are left at the collected
areas, (3) numbers of hairs can be collected evenly across
the entire scalp, and (4) implanted hairs take well.

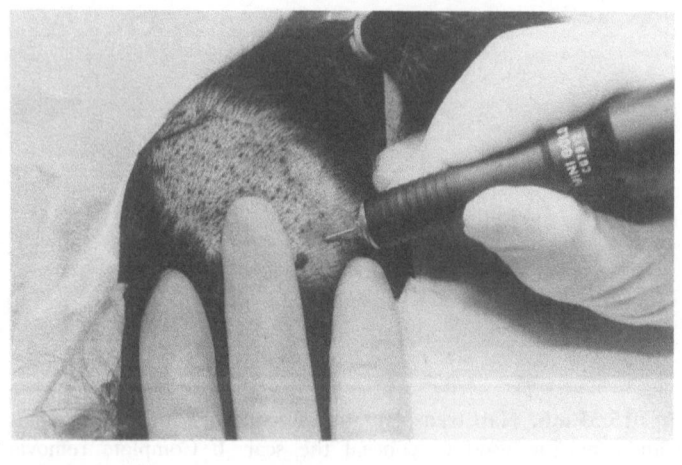

A

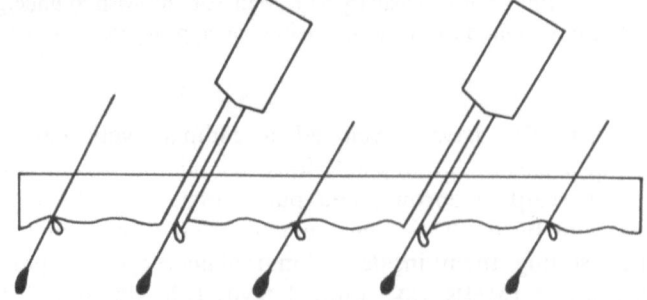

A'

Fig. 15.36. Single hair transplantation method. **A** The micro
punch instrument is used to punch the scalp parallel with each hair
to be collected

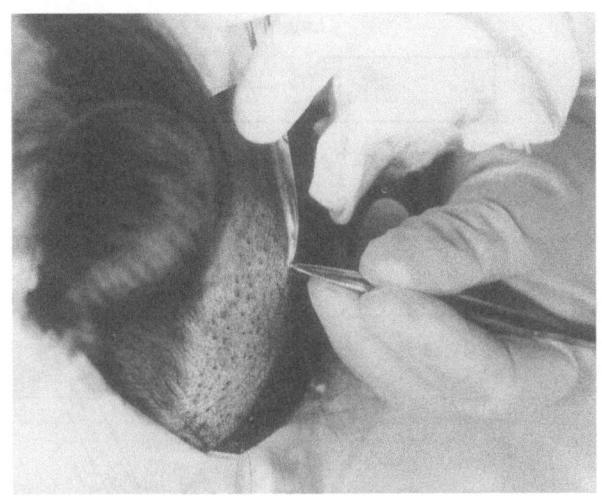

B

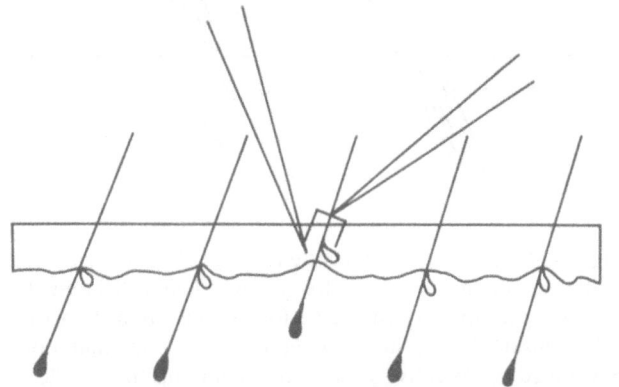

B'

Single
hair

C'

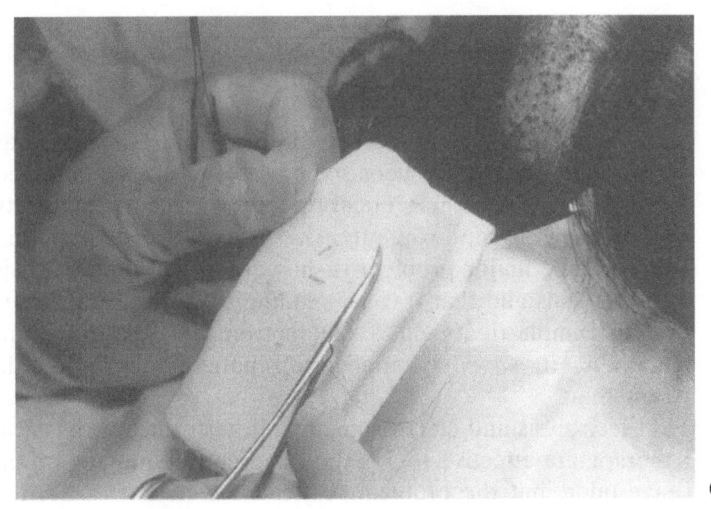

C

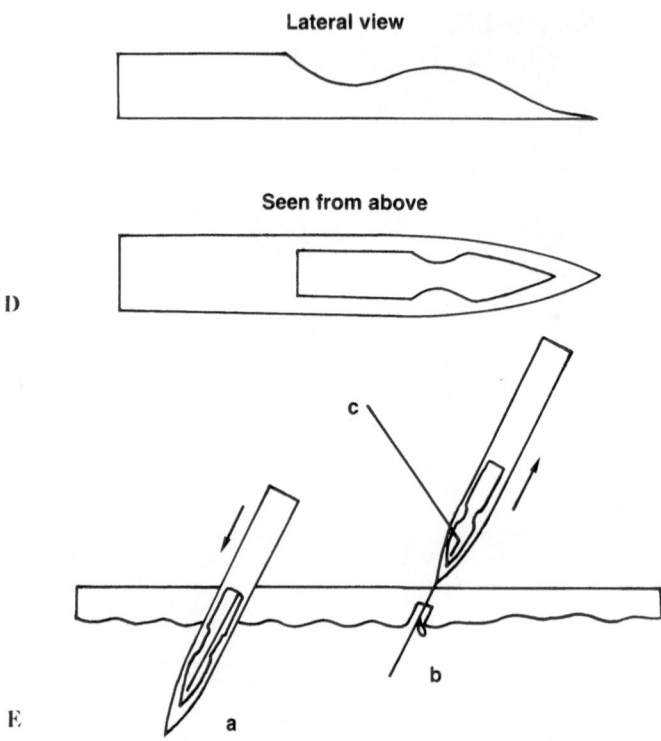

Fig. 15.36. (*contd.*) **B** A pincette is used to hold the punched micro graft and lift the hair bulb for a single hair collection. **C** Single hair. **D** Specially devised needle for hair insertion. **E** A single hair is held inside the specially devised needle (*a*). Implantation can be performed comparatively easily by pulling up the needle (*b*) with the hair being pressed down by a needle (*c*)

15.9.15 Comparison of Efficacy Among Many Types of Treatment

The treatments for bromidrosis and hyperhidrosis as well as their side effects have been explained and are summarized in Table 15.2. Pharmaceutical therapy is effective, but only temporarily, in removing offensive axillary odor to a certain extent. The major problem is, however, that hyperhidrosis still remains and that it causes soiling of clothes. However, as this simple treatment is superb from the aesthetic point of view, it is widely applied on patients showing mild symptoms.

Electrolysis and electrocoagulation performed as physical therapy are effective to a certain degree in removing offensive odor, but the problem related to secretion of eccrine

Table 15.2. Comparison of efficacy among different treatment types

| Treatment | Ease of application | Rest after treatment | Complication | Esthetic result | Results | | Total evaluation |
					Odor persists	Hyperhidrosis persists	
Pharmacy therapy	Excellent	Unnecessary	None	Very good	Some	Yes	Good
Electrolysis	Poor	Unnecessary	None	Very good	Some	Yes	Poor
Electrocoagulation	Poor	Unnecessary	Sometimes	Very good	Some	Yes	Good
Radiation therapy	Excellent	Unnecessary	Always	Very good	Yes	Yes	Poor
Removal therapy	Excellent	Necessary	Usually	Poor	No	No	Poor
Curettage method	Good	Essential	Always	Very good	Yes	Yes	Poor
Clearance method	Good	Essential	Always	Good	No	Yes	Good
Subcutaneous shaver method	Excellent	Unnecessary	None	Very good	No	No	Excellent

(Reproduced with permission from Inaba 1986)

sweat which is remote from hair follicles still remains. The presence of hyperhidrosis after removal of underarm hair is truly a major problem for patients. The complete removal of hair is somewhat defective in that it takes a fairly long period of time for treatment. However, with the development of the latest method, it is recommendable from the esthetic point of view. It is applied only in the case of patients who have developed mild symptoms of hyperhidrosis.

The total removal method as surgical therapy must be avoided, because it is not effective in the complete removal of underarm hair. The curettage method must also be avoided as it is ineffective in the complete removal of sweat glands. The clearance method is comparatively effective on the condition that the subcutaneous tissue is removed to the medium or to the thick split-thickness skin graft level for complete removal of sweat glands. However, it also has some drawbacks in that it is difficult to perform as well as being time-consuming.

As described above, all of the present treatments have both advantages and disadvantages; therefore, some better method was needed. The subcutaneous tissue shaving method is superb from the esthetic point of view, because the removal of subcutaneous tissue can be done to the desired level, varying from thin thickness to full thickness, with a surface incision of no more than 1 cm. After treatment, the double tie-over dressing is adjusted for pressure. This results in the secure arrest of bleeding and in relatively free arm movement. Decompression can be done 1 day after the operation and revascularization occurs on the skin. A satisfactory result is obtained because it helps prevent necrosis.

There had been no reliable radical treatment until this method was developed. It is considered the best method in treatment of bromidrosis and hyperhidrosis. Choi (1988) reported that, using the subcutaneous shaving method, he operated on 1,200 cases of bromidrosis with the following results: the odor disappeared in 1,158 cases (96.5%) of the 1,200 cases, and sweating showed a dramatic decrease in 1,186 cases (98.8%). No side effects or sequelae were noticed with this method of operation.

The more than 20,000 cases treated to date confirm the excellence of this procedure. However, simple as it may seem, the method cannot be performed without thorough training in the various steps of shaving, postoperation dressing, and treatment of various complications both during and after the operation.

Chapter 16.   The Question of
Hair Regeneration

As described previously, physical therapy is used for treatment of axillary bromidrosis by epilation of axillary hair (Chap. 13.3). The fact that hairs frequently regrow in spite of supposedly complete destruction of hair roots has been attributed in many cases to a failure to destroy the hair roots in the telogen stage.

In the subcutaneous tissue shaving method, too, axillary hair is often regenerated even after the lower portion of the hair follicles has been completely removed. This might be construed to be due to remaining hair bulbs in the telogen phase which were situated much higher than the anagen bulbs. Postoperative examination of the skin undersurface, however, reveals that the sebaceous glands have been removed or exposed. This phenomenon is difficult to account for if we subscribe to the classical concept that hair regeneration is based on the persistence of the dermal papilla and the formation of a new bulb through inductive influence of the dermal papilla on follicular epithelium (Chap. 2.5.1, Fig. 2.11).

In treating male patients, we apply rough and medium shaving techniques in order to preserve the sebaceous glands as much as possible. Regeneration of axillary hair in these patients is observed postoperatively even if subcutaneous tissue and dermis are removed almost to the level of the sebaceous glands (Fig. 16.1). For female patients, we perform more complete shaving by adjusting the shaver so that the blade is set at a slightly more upright angle relative to the pressure roller, thus removing the sebaceous glands as well. All or at least most of the sebaceous glands are ablated, so that little more than the infundibulum is preserved. These patients rarely experience subsequent regeneration of axillary hair. Hair is regenerated more often in male patients whose sebaceous glands and follicular

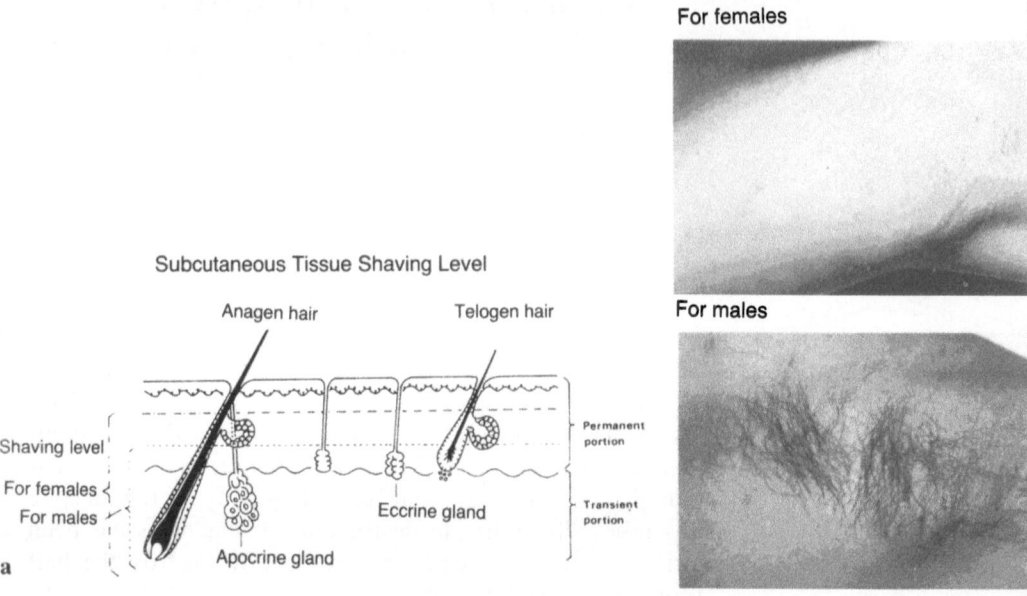

Fig. 16.1.a,b. Ablation of the sebaceous glands prevents regrowth of axillary hair. (Reproduced with permission from Inaba and Inaba 1990)

isthmi were deliberately left intact than in female patients whose sebaceous glands have been completely removed.

16.1 Clinical Findings

We were compelled to question the generally accepted model of the hair cycle after we developed the subcutaneous tissue shaving method for treatment of bromidrosis or hyperhidrosis (Inaba and Ezaki 1977; Inaba et al. 1978c; Inaba 1986).

  Clinical examination confirms that even if the lower part of the hair follicle has been cut off, with only the upper part of the isthmus and the sebaceous glands remaining, regeneration of axillary hair is observed. If we ablate all, or at least some, of the sebaceous glands, so that little more than the infundibulum is preserved, the axillary hair does not regrow (Fig. 16.1). This finding prompted us to attempt histological studies. In this effort we were greatly aided by our development of a special method to prepare thick tissue specimens (Inaba et al. 1978b).

16.2   Thick Tissue Specimen Preparation

The standard thickness of a section of prepared tissue specimen embedded in paraffin blocks is 0.003–0.005 mm. In order to obtain a thick section of 0.1 mm, one would have to "stack" more than 200 of these sections. This is not only troublesome but rather impractical.

The cell tissue is so thin that it curls up and shrinks when each section is sliced. In order to flatten and extend it, the section is placed horizontally in hot water, then laid on a glass surface for staining. However, the tissue structure itself has been affected and has shifted position. Therefore, when these thin sections are put together to create a thick section, it is not possible to study the tissue in its original condition. True, very fine details can be observed with individual thin sections, but with three-dimensional sections we could not observe the stages of the hair cycle as a whole.

It was obvious that some new method of preparing thick tissue specimens had to be worked out (Inaba et al. 1978b). We tried out a number of possibilities. First, we froze the tissue and then sliced it. Since the slicing had to be done by ice-planing by compressed CO_2 gas, however, the tissue specimen would break apart if we tried to slice it thickly. Next we tried a cryostat technique, to make a preparation of 0.04–0.05 mm thickness. But in this instance, even by slicing the tissue at a temperature of $-70°C$, which requires an extremely expensive microtome, the desired thickness could not be obtained. The third effort involved the use of celloidin, in which the tissue was permeated with celloidin and then cut thickly. But it took 2–3 months for the celluloid substance to fully permeate the tissue. Though easy to slice, after waiting that long, details were unclear because the tissue had solidified with the celluloid and staining was inadequate. Each of these methods had its respective advantages and drawbacks, and we eventually came back full circle to the paraffin block, once again attempting thick slicing. When we did this forcefully, the tissue specimen, as usual, curled or broke into fragments somewhat like the surface of a delicate wood board subjected to rough planing.

We then thought of cellotaping the specimen before slicing. Cellotape was attached to a paraffin-embedded specimen (Fig. 16.2.a); with this innovation, the tissue sliced well at 0.2–0.3 mm thicknesses. The thick-section specimen was pressed onto a glass slide by spreading a thin layer of egg albumen glycerine on the center of the slide to prevent the specimen from peeling off (Fig. 16.2.b); then, with the

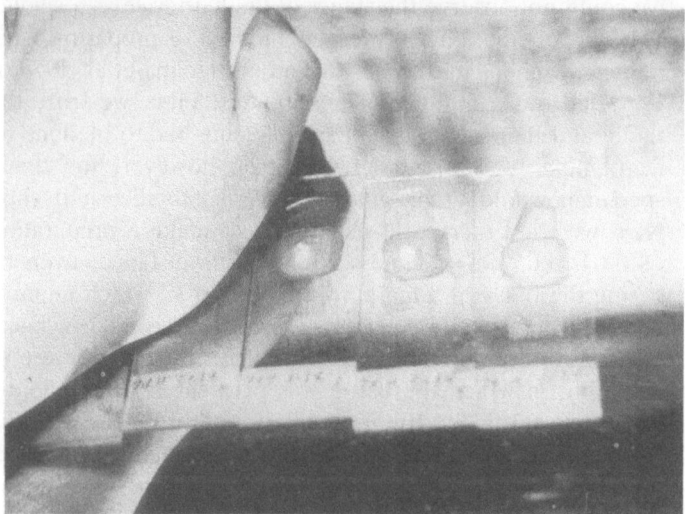

Fig. 16.2.a–c. Preparation of thick tissue specimens using cello-tape. **a** Cellotape is attached to a paraffin-embedded specimen. The latter is then cut to the required thickness (100–200 μm) with a microtome. **b** The thick-section specimen is pressed on to a glass slide. **c** Removal of the cellotape. Soaking the specimen in xylene suffices. The *arrow* indicates the thick specimen

cellotape uppermost, it was firmly pressed with the palm onto the slide. Removal of the cellotape without disfiguring the specimen was made simple by using xylene to remove both cellotape and paraffin at the same time (Fig. 16.2.c).

However, because of the specimen thickness, the paraffin

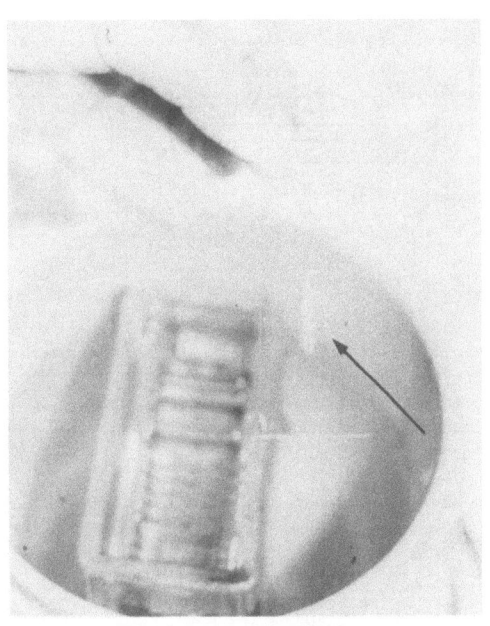

Fig. 16.2.c.

remained in the tissue; staining was still incomplete and left details unclear. To solve that problem, we decided to try immersing the specimen with xylene in warm water (but not by heating, since xylene is flammable) and found that after the cellotape is removed the paraffin could be eluted from the tissue in a xylene-water solution heated at 50°–60°C. Then, to ensure that paraffin removal is complete, the specimen is put in fresh xylene for about 60 min. When we stained the tissue at that time with a double-dilution staining solution, the specimen details were quite clear. Putting together four or five of these thick tissue specimens in serial form, we could readily make the three-dimensional observations we wanted (Figs. 16.3.a,b). The development of this thick tissue preparation method made it possible to observe the actual process of hair regeneration as well as the relationship between sweat glands, hair follicles, and sebaceous glands in three-dimensional clarity (Inaba et al. 1979a).

16.3 Common Hair Cycle

The common model of the hair cycle is divided into three stages: anagen, catagen, and telogen. In the anagen stage, the new hair is formed and grows rapidly, with vigorous

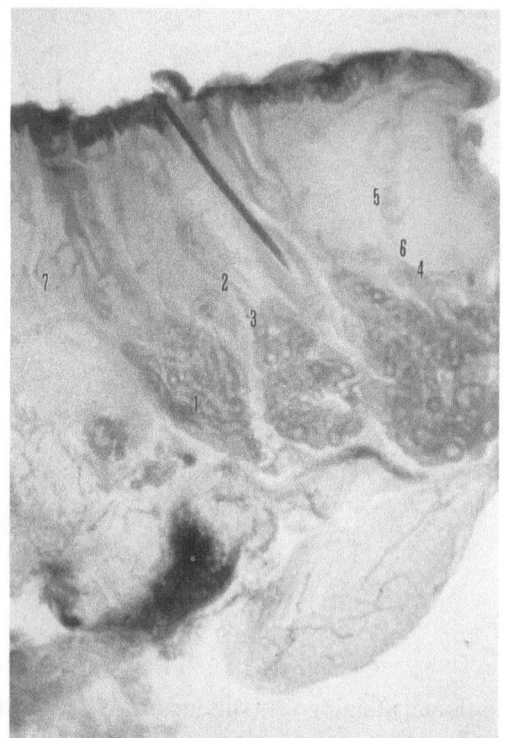

Epidermis

Dermis

Subcutaneous
layer

a

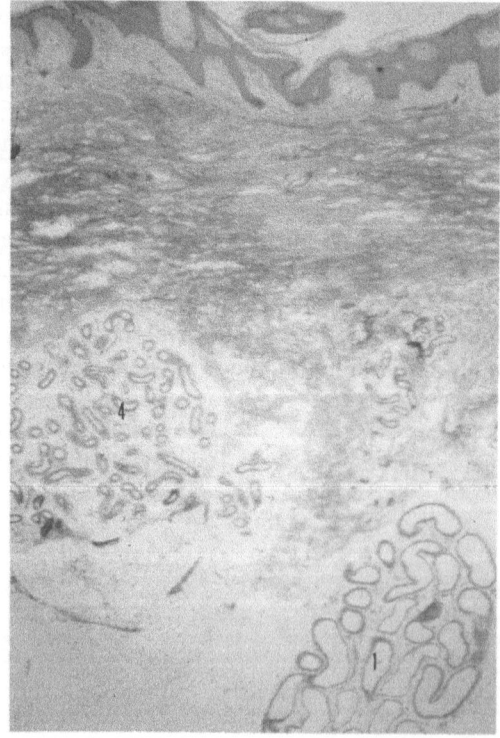

b

follicular activity as well as much mitotic activity in the bulb matrix (Fig. 2.11, Chap. 2.5.1). In the catagen stage, the hair follicle has ceased to produce a new hair, which assumes a broomlike configuration and becomes a club hair. In the telogen stage, the follicle ceases its reactivity, and the length of the follicle is now one-third to one-half shorter than it was in the anagen stage. From this time, the lower portion of the follicle moves upward to a point close to the arrector pili muscle and the hair root becomes superficial, leading to loss of the expiring hair (telogen stage).

According to the common theoretical model, the hair bud (germ) is formed at the lower tip of the follicle in the telogen stage (Figs. 16.4.a–c). The bud then descends downward at a certain angle to form the hair bulb. This is called the hair germ stage. After the bulb is fully developed, new hair tissue is formed from mitotic activity in the matrix. This tissue makes up the initial hair cone which later corresponds to the inner root sheath. The new hair itself is then formed within the cone (inner root sheath). This is the hair peg stage. The new hair bulb moves downward in a synchronous fashion to a point which becomes the lower tip of the anagen stage hair follicle (bulbous peg stage) (Fig. 2.11). Thus, when the new hair begins to grow upward, the preexisting hair, now in the telogen stage, is pushed out and discarded.

The common model of the hair cycle states, in essence, that the new hair bud is formed from the lower tip of the telogen follicle and that the hair itself is generated from the matrix after full formation of the hair bulb. In other words, it has been thought that a hair cannot be formed unless and until a hair bulb and matrix are already present.

16.4 Histological Findings

Histological studies (Inaba et al. 1978b; Inaba 1988; Inaba et al. 1988d) using these specially prepared thick tissue specimens have revealed that when the sebaceous gland is left intact, the new young bud (hair germ) begins to form at

Fig. 16.3.a,b. Comparison of conventional and cellulose tape methods of thick tissue preparation, **a** Conventional thin-thickness tissue preparation (3–5 μm thickness). **b** Thick tissue preparation (150 μm). Three-dimensional observation is now possible. 1, Apocrine gland; 2, Apocrine duct; 3, Apocrine coiled duct; 4, Eccrine gland; 5, Eccrine duct; 6, Eccrine coiled duct; 7, Sebaceous gland

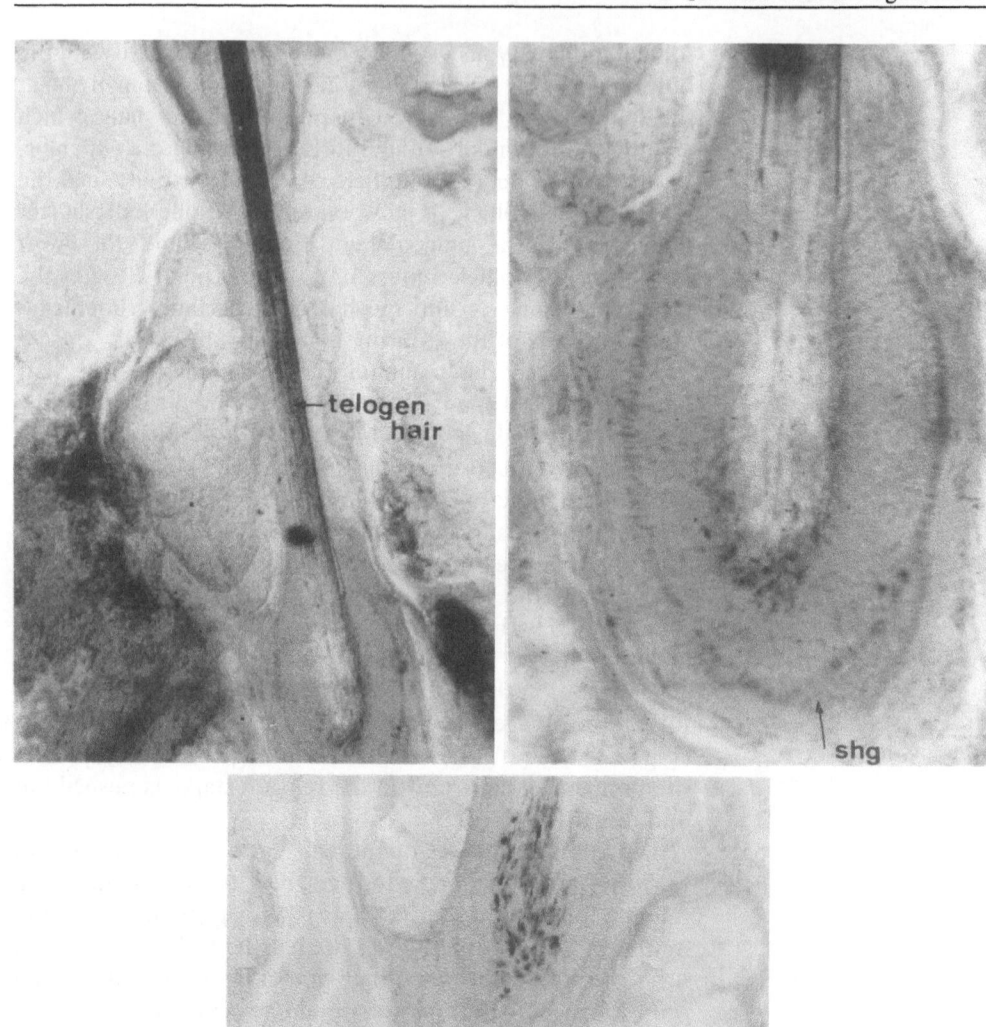

Fig. 16.4.a–c. The common hair cycle. **a** Telogen hair follicle.
b According to the common hair cycle, the portion indicated by
the *arrow* is the secondary hair germ (*shg*) of the telogen follicle.
c The new young hair germ (*arrow*) descends from the lower
portion of the telogen follicle

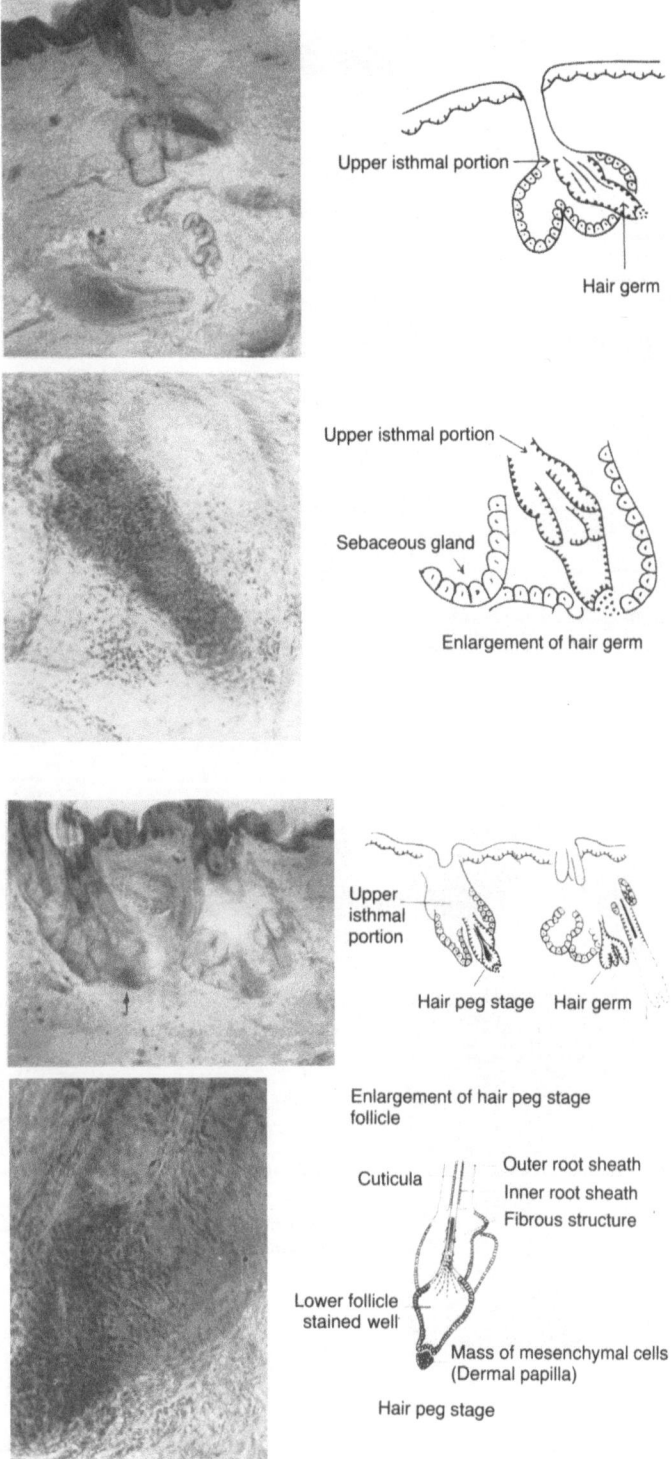

Fig. 16.5.

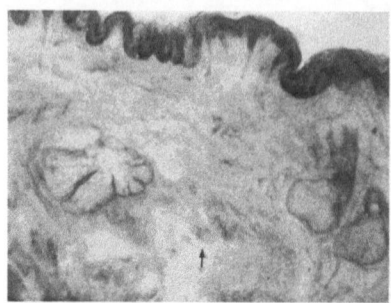

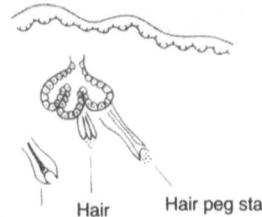

Terminal   Hair   Hair peg stage
hair   germ stage

c

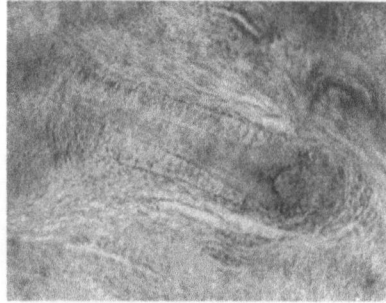

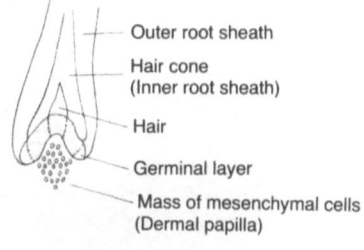

Outer root sheath

Hair cone
(Inner root sheath)

Hair

Germinal layer

Mass of mesenchymal cells
(Dermal papilla)

Hair peg stage

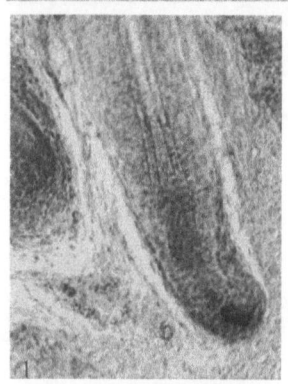

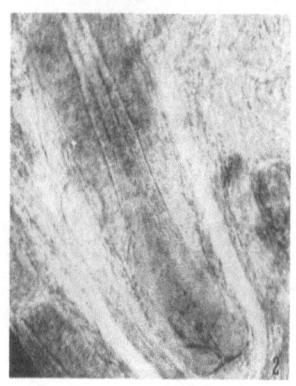

Formation of the hair matrix

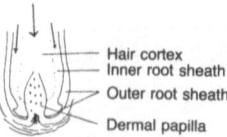

Hair cortex
Inner root sheath
Outer root sheath

Dermal papilla

**Schematic representation of early events in the
replacement of vellus hairs by coarse hairs**

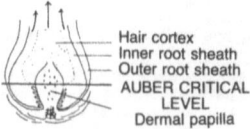

Hair cortex
Inner root sheath
Outer root sheath
AUBER CRITICAL
LEVEL
Dermal papilla

**Schematic representation of later events in the
replacement of vellus hairs by coarse hairs**

d

a point adjacent to the duct opening of the sebaceous gland (upper isthmal portion). This hair germ becomes multilobular and forms a bud (Fig. 16.5.a). Although the hair bulb (matrix) has not yet developed, filamentous structures are produced from the surrounding germinal layer in the hair germ, and then keratinized to form the inner root sheath (Henle's layer, Huxley's layer, and sheath cuticle) and new young hair (hair cuticle, cortex, and medulla) (Figs. 16.5.b, 16.6, hair germ stage).

As the hair bud elongates downward, this germinal layer becomes localized at the lower portion of the hair follicle above the dermal papilla. Although the hair bulb is not yet formed at the hair peg stage, the lower germinal layer is already divided into three parts: the outer root sheath, the inner root sheath and the hair tissue (hair peg stage) (Figs. 16.5.c, 16.6). As the hair follicle begins to peg downward, the inner root sheath elongates downward. The germinal layer wraps around a mass of mesenchymal cells to form the matrix and hair bulb (bulbous peg stage) (Figs. 16.5.d, 16.6).

Since the new hair is prevented from growing upward by firm interlocking fusion of the cuticle, the follicle continues to grow downward with vigorous mitotic activity in the newly formed matrix. In this bulbous peg stage, however, the formation of the matrix is still incomplete and not yet compressed as it is in the terminal hair stage.

When the forming hair follicle reaches a level near the permanent site of the dermal papilla, the matrix grows around the mass of mesenchymal tissue, thereby pressing the outer root sheath outward. Consequently, the outer root sheath is thinned, and has only one or two layers at its bottom. It is evident that the germinal layer surrounding the dermal papilla may be divided into six distinct regions:

Fig. 16.5.a–d. The formation of new hair buds. **a** Hair germ stage: the hair germ begins to form from the duct opening of the sebaceous gland (upper isthmal portion). **b** Hair peg stage (*arrow*): although the hair bulb (matrix) has not yet formed, filamentous structures are formed in the surrounding germinal layer and then keratinized to form the hair shaft. **c** Hair peg stage (*arrow*): as the hair bud elongates downward, this germinal layer becomes localized at the lower portion of the hair follicle above the dermal papilla. **d** Bulbous peg stage: the inner root sheath elongates downward to form the hair bulb. The germinal layer wraps around a mass of mesenchymal cells to form the matrix (*inserts 1 and 2*). The matrix grows, pressing the outer root sheath to form the bulb (*insert 3*)

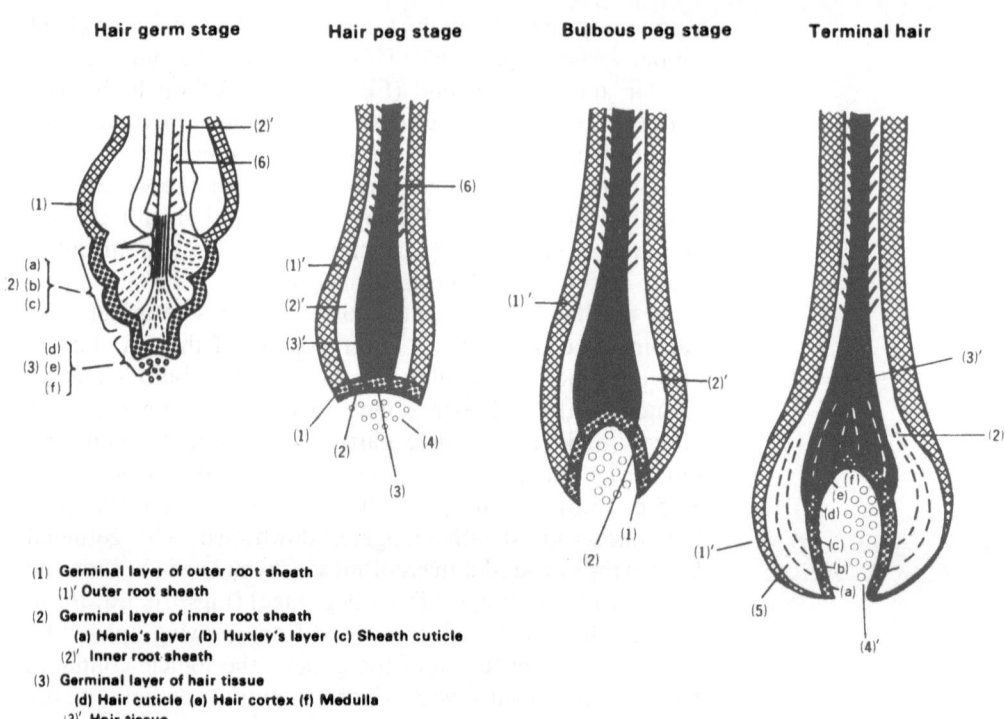

| Hair germ stage | Hair peg stage | Bulbous peg stage | Terminal hair |

(1) **Germinal layer of outer root sheath**
    (1)′ **Outer root sheath**
(2) **Germinal layer of inner root sheath**
    (a) **Henle's layer** (b) **Huxley's layer** (c) **Sheath cuticle**
    (2)′ **Inner root sheath**
(3) **Germinal layer of hair tissue**
    (d) **Hair cuticle** (e) **Hair cortex** (f) **Medulla**
    (3)′ **Hair tissue**
(4) **Mesenchymal cells**   (4)′ **(Dermal papilla)**
(5) **Matrix**
(6) **Cuticula**

Fig. 16.6. The process of hair follicle formation after subcutaneous tissue shaving. The hair germ begins to form at a point adjacent to the duct opening of the sebaceous gland in regeneration of the hair follicle. Although the hair bulb has not yet formed, the inner root sheath and new young hair are already formed from the surrounding germinal layer. (Reproduced with permission from Inaba and Inaba 1990)

Henle's layer, Huxley's layer, sheath cuticle, hair cuticle, hair cortex, and hair medulla (Fig. 16.6). As the hair bulb is formed, the hair and inner root sheath grow at the same speed upward toward the skin surface, and the terminal hair then grows according to the classic theories of the hair cycle (Figs. 16.5.d, 16.6, terminal hair).

This finding has given us information of clinical value in setting our shaving instrument to assure regrowth or no regrowth of axillary hair, as the patient prefers. It has far more importance, of course, in defining a more accurate model of the human hair cycle.

In the process of hair regeneration, hair germs are formed at the portion of the hair follicle adjacent to the excretory duct opening of the sebaceous glands (upper isthmal portion) and then continue to grow, going through the hair peg and bulbous peg stages, finally to become terminal hairs. This finding was supported by several studies (Inaba et al. 1981. Inaba et al. 1982, Inaba 1985).

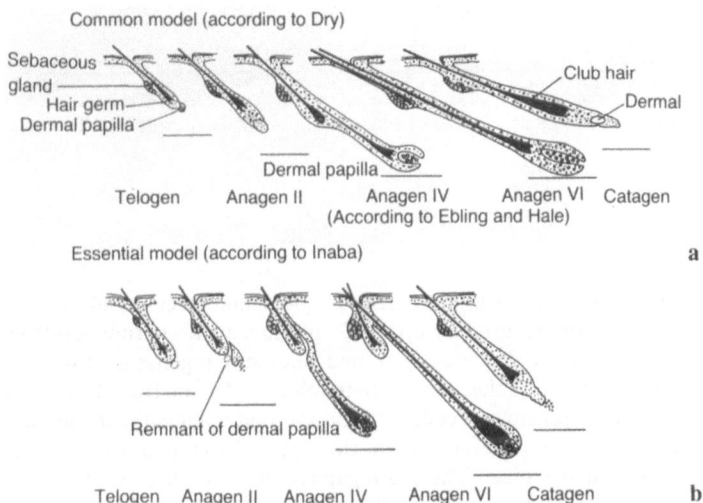

Common model (according to Dry)

Sebaceous gland

Hair germ

Dermal papilla

Dermal papilla

Club hair

Dermal

Telogen   Anagen II   Anagen IV   Anagen VI   Catagen
(According to Ebling and Hale)

Essential model (according to Inaba)

a

Remnant of dermal papilla

Telogen   Anagen II   Anagen IV   Anagen VI   Catagen

b

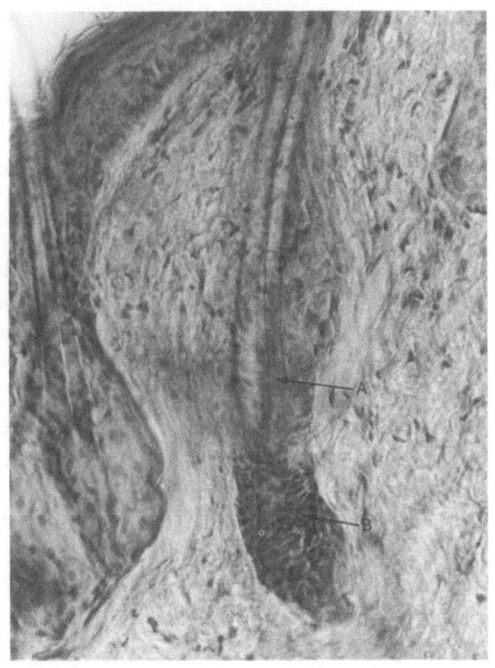

c

Fig. 16.7.

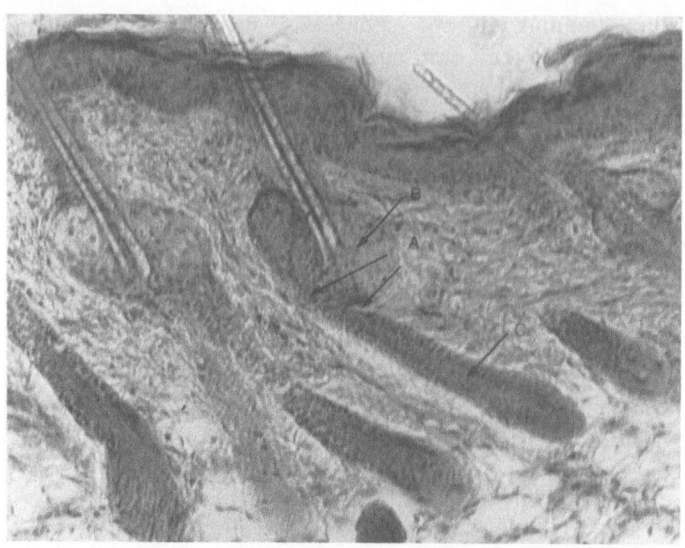

d

Fig. 16.7.a–d. a Common (Dry) model and **b** essential (Inaba)
model of the mouse hair cycle. (Reproduced with permission from
Inaba 1985). **c** The new hair bud does not regenerate from the
lower portion of the telogen hair follicle (*A*) but from the newly
formed mesenchymal cells (*B*) and the epithelial cells. **d** The new
hair follicle regenerates from the upper isthmal portion of the
telogen hair follicle. There is a gap (*A*) between the lower portion
of the telogen hair follicle (*B*) and the new young hair follicle (*C*)

16.5   Hair Regeneration in the Mouse

We attempted to study the process of regeneration by
applying an anti-cancer agent (methotrexate, etc.) to mouse
body skin surface, thus experimentally creating telogen hair
follicles (Inaba 1985). The new hair bud (germ) in the
regeneration process has been thought to begin to form
from remnants of the dermal papilla or a secondary hair
germ in telogen stage (Fig. 16.7.a). However, the authors
recognize from the recent studies that the regeneration of
the hair germ does not necessarily occur from the remnants
of the dermal papilla, but can be formed from the newly
formed mesenchymal cells (Figs. 16.7.b,c). This finding also
indicates that this regeneration is closely related with the
isthmal portion and the vasculation of the hair follicle
(Inaba 1985) (Fig. 16.7.d).

The common hair cycle theory states that during the telogen stage, hair regeneration starts from the lower end of the compound hair follicle where 2–3 hairs are present in a single follicle (Fig. 16.8.a). Contrary to this, our finding is that a single mass of mesenchymal cells is sometimes newly formed at the lower end of compound hair follicles above which the epithelial cell layer will migrate (Figs. 16.8.b(i),d(i)). It suggests that generation of the mass of mesenchymal cells does not necessarily start from the remnant of dermal papilla cells or secondary hair germ present at the lower portion of telogen follicles, but the recovery of the vascular system surrounding the follicles may precede the above phenomenon.

If the vascular system or the upper isthmal portion of the sebaceous gland is considerably damaged, hair is lost permanently as there are no epidermal stem cells present (Figs. 16.8.c(ii),d(iv)). When the damage is only slight and recovery of the vascular system can be observed as a result, a mass of mesenchymal cells starts to generate from the lower end of the follicles. If the telogen hair has been epilated, hair generation is observed to start from the upper

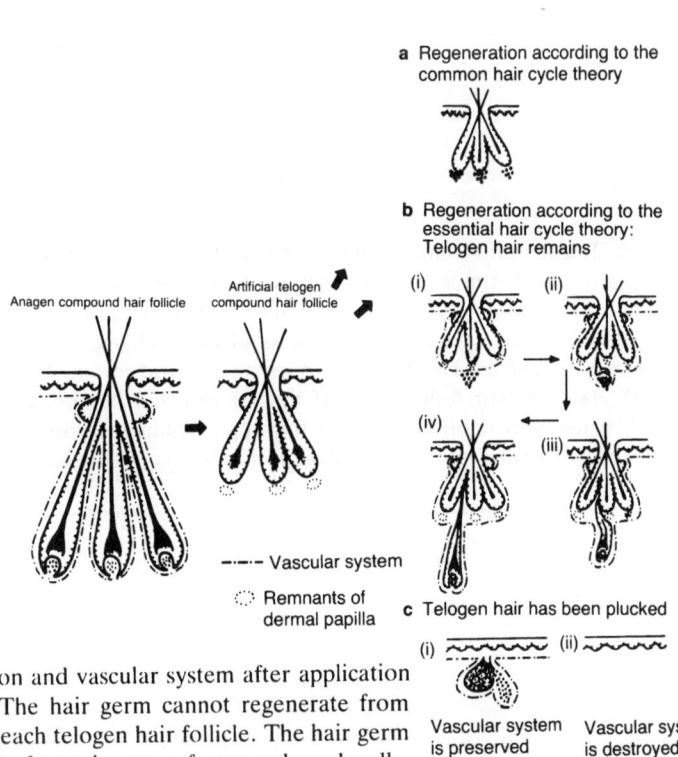

Fig. 16.8.a–c. Regeneration and vascular system after application of the anti-cancer agent. The hair germ cannot regenerate from remnant dermal papilla of each telogen hair follicle. The hair germ is formed from the newly formed mass of mesenchymal cells. (Reproduced with permission from Inaba 1990)

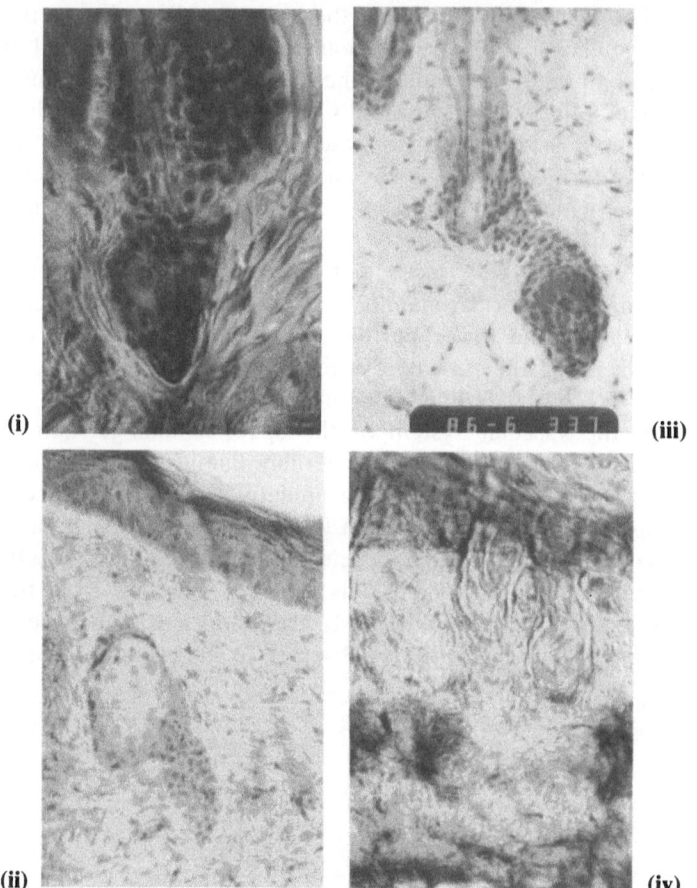

(i)

(iii)

(ii)

(iv)

Fig. 16.8. Regeneration and vascular system after using an anti-cancer agent. (**i**) A single mass of mesenchymal cells was newly formed at the lower portion of the compound hair follicle. (**ii**) The epithelial cell layer migrates above a mass of mesenchymal cells and becomes a matrix. This hair bud does not elongate from the tip of the telogen hair follicle. (**iii**) If the telogen hair has been plucked, generation is observed from the upper isthmal portion. (**iv**) When the vascular system is damaged, hair loss is permanent

isthmal portion close to the duct opening of the sebaceous gland (Figs. 16.8.c(i),d(iii)).

The chance of hair regeneration depends on whether the vascular system of the pilosebaceous unit as well as the upper isthmal portion of the sebaceous gland are left intact (Inaba and Inaba, 1990).

16.6   Vascular System for Hair Follicle Pilosebaceous Unit

Up to the present, it was commonly believed that transverse branches of blood vessels in the cutaneous blood plexus and candelabra vessels in the lower blood plexus supply nutrition to the dermal papilla of hair follicle. Formation of a continuous vascular system for the hair follicle and sebaceous gland appended to it is observed, and this perifollicular network of blood vessels cross-shunt with the upper dermal vasculature, whereas blood is supplied originally from the deep dermis, subcutaneous adipose tissue (Fig. 16.9.a).

Such versatility of the blood supply for the hair follicles is observed specifically during various stages of the hair growth cycles (Montagna and Ellis, 1957). In an attempt to observe blood supply in the pilosebaceous unit, we (Inaba et al. 1987b) ligated a point above the secretory duct opening of the sebaceous gland and the infundibular portion of hair follicles with a fine nylon ligature. Signs of regressive

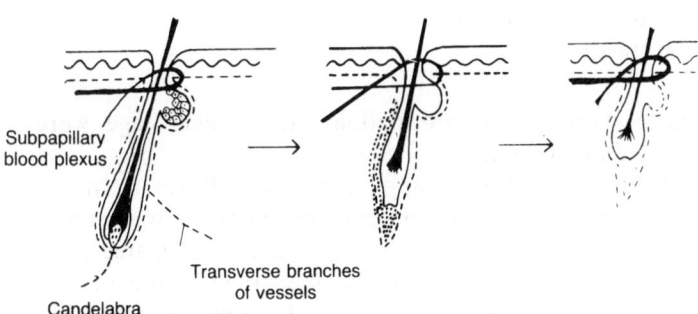

Fig. 16.9.a–c. Changes in hair follicle after binding at upper portion of hair follicle. (Reproduced with permission from Inaba 1990).

Subpapillary blood plexus

Candelabra vessel

Transverse branches of vessels

(i) The blood supply within the pilosebaceous unit is considered to form from cutaneous blood plexuses and musculocutaneous arteries

(ii) If the hair follicle is ligated at the infundibular portion, the hair follicle becomes a catagen follicle with blood congestion

(iii) The hair follicle goes into the telogen stage. This finding indicates that there is a vascular system (microcirculation), forming a continuous unit from the subpapillary plexuses   **a**

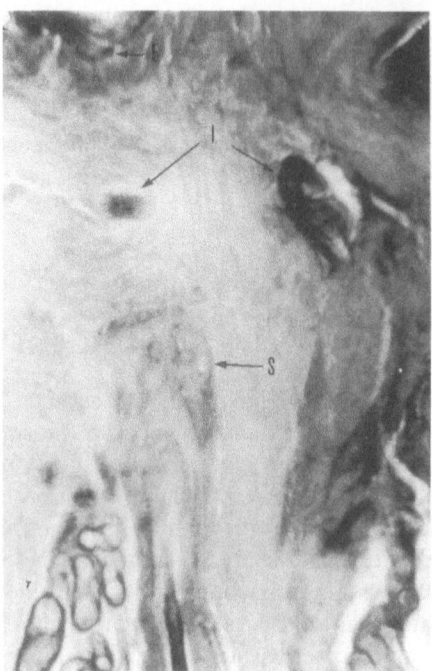

b(i)

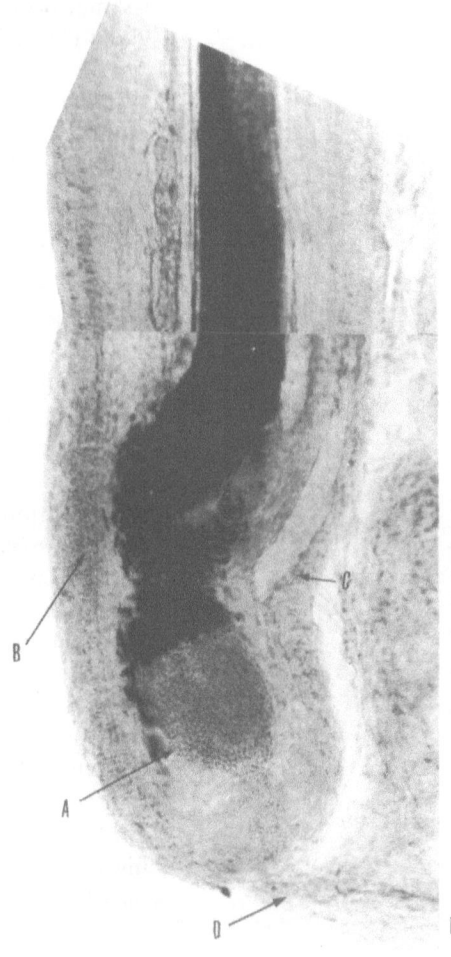

b

Fig. 16.9.b. (i) The infundibular portion of the hair follicle is ligated with a fine nylon ligature. L, ligature; S, atrophied sebaceous gland; E, epidermis. (ii) The lower portion of the hair follicle with blood congestion. A, blood congestion is observed inside the dermal papilla; B, blood congestion is observed inside the left portion of the connective sheath; C, less blood congestion is observed inside the right portion of the connective sheath; D, no congestion is observed in the candelabra vessel or the transverse branches of vessels.

degeneration and transformation into telogen stage were observed in all the hair follicles (Fig. 16.9.a).

Although blood was supplied to the follicles from the vascular system in the hypodermal plexus (Fig. 16.9.a(i)), if the infundibular portion of the hair follicle was ligated (Fig. 16.9.b(i)), blood congestion was seen only in the dermal papilla and the connective sheath and not in the candelabra vessel or the transverse branch vessels (Fig. 16.9.b(ii)). Figure 16.9.c indicates that if the ligation has been done as described above, the initial catagen stage is induced. This finding indicated that a microcirculatory system starting from the subpapillary blood plexus and surrounding the follicular appendix is present and that this vascular system

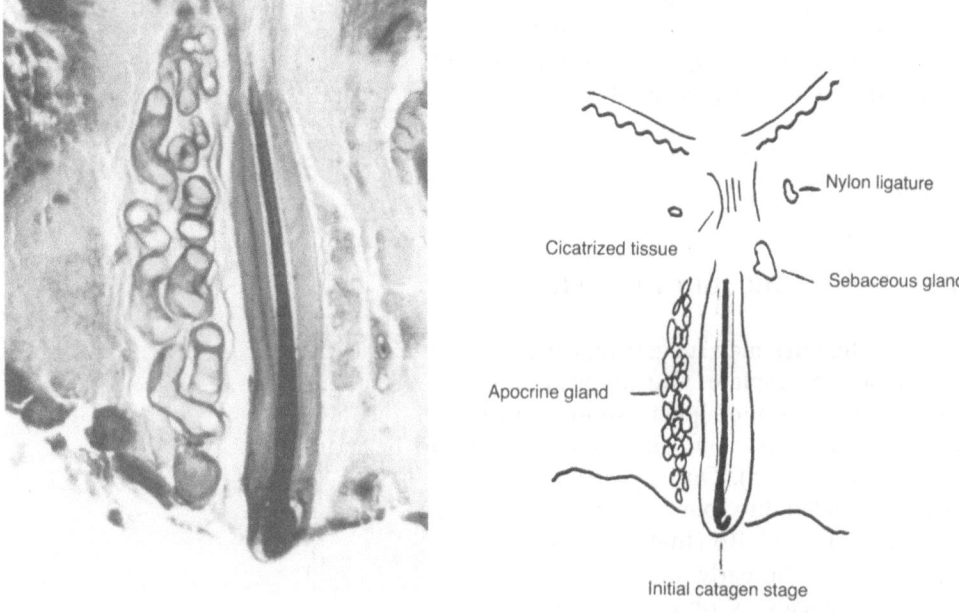

Nylon ligature

Cicatrized tissue

Sebaceous gland

Apocrine gland

Initial catagen stage

(ii)

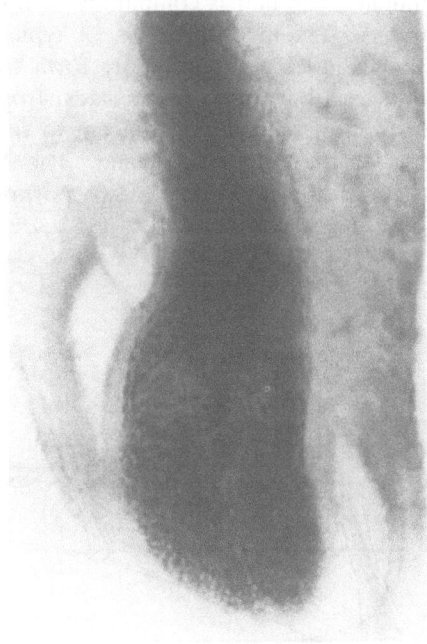

Fig. 16.9.c. (i) Initial catagen stage after binding at the upper portion of the hair follicle. (ii) Diagram of initial catagen stage. (iii) Enlarged view of the lower portion of the hair follicle

(iii)

is of prime importance, with the principal blood supply coming from the subpapillary blood to form the micro-circulation around the pilosebaceous unit. The blood vessels supplied from the dermal arterial plexus are only of second-ary importance. From this finding, it can be surmised that, in the sebaceous gland, enzyme and hormone activities are continuously taking place.

16.7   New Concept of the Hair Cycle

The above findings may suggest that the common hair cycle theory is still incomplete. Our proposition that the true hair center is sited at the upper isthmal portion of the hair follicle and the sebaceous gland then appears to be more reliable (Inaba, 1985). Our suggestion is that the hair cycle should be divided into four stages: anagen, catagen, telogen and isthmal (Fig. 16.10) (Inaba, 1985).

Formation of the early anagen stage from the lower end of telogen follicles as observed in Japanese monkey and human scalp hair is compatible with the common hair cycle theory. Yet in observation of typical telogen hairs, early anagen does not necessarily form from the lower telogen follicles, but also regenerates from the upper isthmal portion of the follicle adjacent to the secretory duct open-ing of the sebaceous gland. The hair follicle has been theoretically divided into two portions, transient and per-

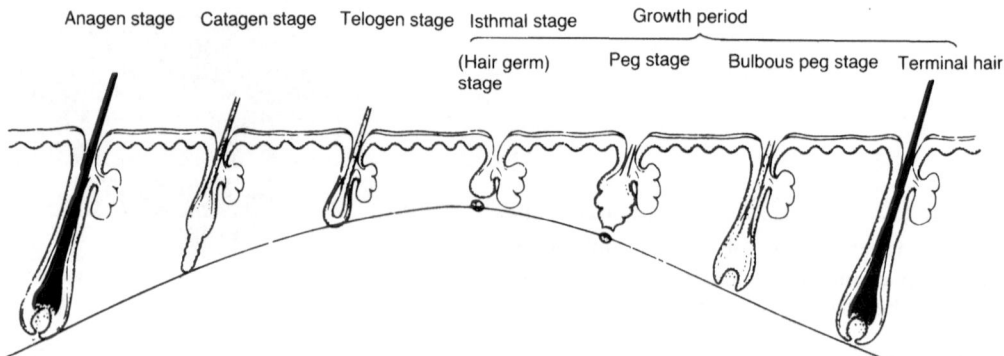

Fig. 16.10. The essential hair cycle. The authors suggest that an isthmal stage should be added to the conventional hair cycle since it has been found that when telogen hair is epilated, regene-ration has been observed from the upper isthmal portion of the hair follicle at the secretory duct opening of the sebaceous gland. (Reproduced with permission from Inaba and Inaba 1990)

manent, at the lower telogen hair follicle's point of attachment to the arrector muscle (Montagna and Parakkal, 1974) (Fig. 16.11). The common hair cycle theory would seem valid if regeneration were to start from this remnant of dermal papilla cells or secondary hair germ which is present at the lower telogen follicle sac. The regeneration from the hair germ depends on the size of the telogen hair because, in some cases, the telogen hair follicle retracts within the lobes of the sebaceous gland instead of extending to the site of attachment of the arrector pili muscle. We found specifically that if the telogen hair is plucked, regeneration can be observed to start from the upper isthmal portion of the hair follicle adjacent to the secretory duct opening of the sebaceous gland (Inaba and Inaba 1990).

The point of regeneration in our essential hair cycle theory is thus sited at the upper isthmal portion of the telogen hair follicle (essential hair cycle) (Figs. 16.10, 2.14.a,b).

16.8   Essential Hair Cycle Hypothesis and Supportive Findings

Montagna and Parakkal (1974) separated the hair follicle into two parts: the permanent and transient portions. Pinkus (1969) proposed to separate it into three parts: the upper portion (from the epidermis to the duct-opening site of the sebaceous gland), the middle portion (from the duct-opening site of the sebaceous gland to the lower portion of

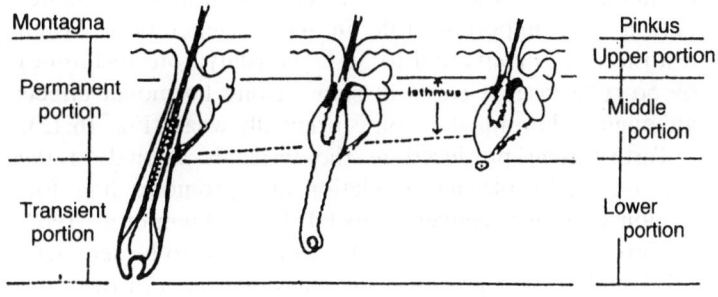

Anagen stage   Catagen stage   Telogen stage

Fig. 16.11. The hair cycle. *Left*, anagen; *center*, catagen; *right*, telogen. Terminology used by Montagna and Parakkal (1974) and Pinkus (1969) for the different portions of the hair follicle are shown in the margins. (Reproduced with permission from Inaba et al. 1979a)

telogen phase), and the lower portion (the same as the transient portion) (Fig. 16.11).

The central portion of the hair follicle has been said to lie at the lower portion of the telogen stage follicle. It is now evident that this central portion is instead situated exactly at the isthmal portion close to the duct opening of the sebaceous gland. The hair bud can be formed from the isthmal portion. The new young hair is already formed at the lower portion of the germinal layer prior to formation of a new hair bulb.

Pinkus and Mehregan (1981) cited our hypothesis (Inaba et al. 1979a) in their book (*The Guide to Dermatohistopathology 3rd edn*): "This general scheme of the hair cycle has recently been challenged by Inaba et al. who found that axillary hair of Japanese is newly formed in the region of the sebaceous duct when dermal papilla and the entire lower part of the hair follicle have been surgically removed."

16.9 Sebaceous Gland Hypothesis for Male Pattern Baldness

If the above basic concepts prove to be incorrect, the causative mechanism will not be revealed and effective treatment will not be developed for male pattern baldness.

16.9.1 Review of the Hormonal Mechanisms

Hamilton (1942) was the first to report that male hormones play a very important role in hair regeneration. When testosterone is converted to 5α-DHT (dihydrotestosterone) by 5α-reductase, it starts to show strong hormonal effects although its hormonal action is originally weak (Fig. 16.12).

Testosterone produced in the testes is carried to the hair matrix by plasma circulation and promotes hair formation when it is converted to DHT (Dorfman et al. 1962, Anderson and Fulton, 1973). It still remains to be seen why DHT inhibits hair growth in the parietal region of the scalp while, on the other hand, it promotes the growth of beard, chest hair, or axillary hair. Adachi and Kano (1970) found that DHT inhibits the adenyl cyclase activity and cell proliferation is inhibited as a result. The apparently contradictory phenomenon of 5α-DHT both promoting growth of beard, chest hair or axillary hair and at the same time inhibiting that of hair in the parietal region has yet to be

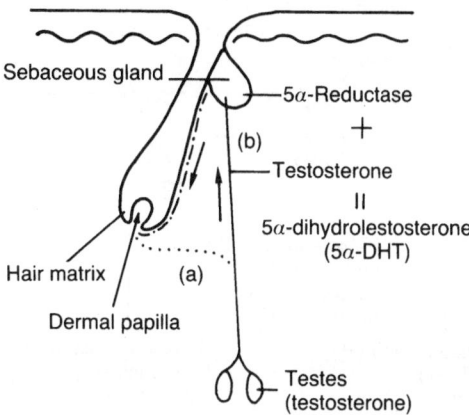

Fig. 16.12. Circulation of testosterone. Testosterone is converted to DHT mainly by 5α-reductase in the sebaceous gland, then binds to the receptors in the sebaceous gland cells, and acts on the nucleus to promote proliferation and enlargement of cells. This DHT acts secondarily on the dermal papilla and the matrix. (Reproduced with permission from Inaba 1990)

fully elucidated, and it has generally been accepted that hormonal imbalance plays a vital part in it. We (Inaba et al. 1988a) conducted studies on the difference in male and female hormone receptors among various regions and by age but could not find any difference in the results. According to Schweikert and Wilson (1974), more 5α-reductase is present in hair follicles in bald regions than in those of other regions. Although there have been several such reports, the causative mechanism of male pattern baldness has not been clearly elucidated.

16.9.2 Sebaceous Gland Hypothesis for Male Pattern Baldness and Supportive Findings

The process by which male pattern baldness develops can be explained by the sebaceous gland hypothesis, which proposes the true hair center as located in the sebaceous gland and upper isthmal portion. Testosterone is first converted to DHT mainly by 5α-reductase in the sebaceous gland. It then binds to sebaceous gland cell receptors and acts on the nucleus to promote cell proliferation and enlargement. It may also act secondarily on the dermal papilla and the hair matrix through a vascular system which exists in the connective hair follicle surrounding the hair follicles. In brief, the condition of the sebaceous gland affects the hair

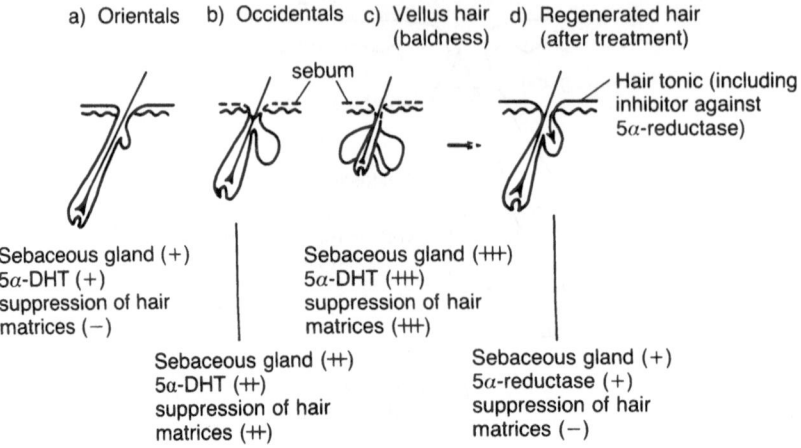

Fig. 16.13. Hair follicles and sebaceous glands relative to 5α-DHT production. Orientals have follicles with quite small sebaceous glands and relatively low incidence of premature male pattern baldness (MPB). Occidentals, who typically consume much greater volumes of animal fats, have comparatively large sebaceous glands and a relatively high incidence of premature MPB. The vellus hair follicle typical of MPB has a supra-enlarged multilobular sebaceous gland, whereas the follicle of the terminal hair regenerated after treatment has a sebaceous gland of quite modest size. Levels of 5α-DHT production tend to be directly proportional to sebaceous gland size. (Reproduced with permission from Inaba 1990)

root (Fig. 16.13). Takayasu et al. (1980) in effect substantiated this finding by reporting that 5α-reductase is detected in greater amount in sebaceous glands and apocrine glands than in the hair matrix. Sawaya et al. (1988) more recently found that sebaceous glands in the bald areas of male pattern alopecia have greater 3β-HSD (Δ^5-3β—hydroxysteroid dehydrogenase) activity converting dehydroepiandrosterone (DHA) to Δ^4-androstenedione (Ad) as well as androstanedion to testosterone and increased androgen binding capacity more so than in hairy scalp areas (Sawaya et al. 1989). These findings may altogether provide a biochemical explanation for the disease mechanism of androgenic alopecia.

As Lucky (1988) points out, the Sawaya findings present another step forward in deciphering metabolic differences between "balding" and "hairy" scalp skin. The enzyme (3β-HSD), in converting dehydroepiandrosterone (DHA) to Δ^4-androstenedione (Ad) and androstanedione to testosterone (T), shows more activity in balding than in hairy scalp samples. By adding the precursor androgen [3H]DHA to isolated sebocytes, they measured the amounts produced by more potent androgenic metabolites, [5H]A and [3H]T. Moreover, they demonstrated that this dichotomous activity was reflected in the specific 3β-HSD activity of the cytosolic fraction of sebocyte homogenates. The Sawaya study also illustrates the phenomenon of sebocytes metabolizing the "weak" androgen DHA to far more potent metabolites, such as T. The differences shown between balding and hairy skin indicate a potential role of excessive DHA in promoting baldness.

16.9.3 Hair Follicles and Sebaceous Glands Relative to 5α-DHT Production

The larger the sebaceous gland becomes, the smaller the hair follicle (Montagna and Parakkal 1974). An abnormally enlarged sebaceous gland overproduces 5α-DHT and this inhibits the hair matrix. The result is that mature terminal hairs become downy (vellus) hairs. In the author's own hypothetical model, this is the principal causative mechanism of male pattern baldness (Inaba 1985).

In the healthy hair follicle, the appended sebaceous gland is not large, so that the volume of 5α-reductase within it is normal. The effect of this enzyme on testosterone does not cause formation of 5α-DHT in large volume. Rather, the formation of 5α-DHT in proper quantities enhances hair growth (Fig. 16.13.a).

If, however, the sebaceous gland is enlarged for some reason, such as genetic inheritance, dietary habits, hormone influence, etc., the quantity of 5α-reductase will also increase and produce more 5α-DHT. Just as excessive use of fertilizer will stunt rather than enhance the growth of plants, too much 5α-DHT can act to inhibit hair growth. The authors have found that the effect of excessive 5α-DHT on the hair matrix is an inhibition of cell division and gradual atrophy of the hair follicle (Fig. 16.13.b). This triggers off the onset of baldness.

The common belief is that baldness occurs as a result of expiration of the hair root. The hair does not regenerate because it no longer has a root. I have said that an atrophied hair follicle is the cause of baldness, but this is not quite the same as loss of the hair root. In fact, the root remains but the hair grows thin and short, as a vellus hair (Fig. 16.13.c).

Likewise, baldness does not imply that the sebaceous gland disappears. The hair can regenerate as long as the hair follicle exists, as I have already noted in touching on my treatment procedure for relief of bromidrosis. My own conviction that baldness can be prevented and cured comes from this particular discovery.

Actual treatment techniques for relief of baldness will be described elsewhere. In brief, however, the key factor seems to be inhibition of excessive 5α-reductase activity in the sebaceous gland, and this provides the chief lead for development of truly effective remedies.

Common premature baldness can be subject to effective remedy. This type of baldness is, of course, quite different from the so-called scar-tissue (alopecia cicatrisata) type which will be described elsewhere. We can find, in the common type, vellus-like hairs growing where the so-called bald spots appear. This fact points to over-production of 5α-DHT and consequent inactivation of the hair follicle. Accordingly, the hair could be theoretically regenerated by inhibiting the secretion of 5α-reductase from the sebaceous gland (Fig. 16.13.d).

Once again, premature baldness (M.P.B.) is not a phenomenon in which the hair root and sebaceous gland have expired. The hair follicle becomes less active and atrophies, but does not disappear. Regeneration of terminal hair, with proper remedial treatment, is not impossible.

16.9.4   Conclusion

With the development of the radical treatment for bromidrosis, we were able to elucidate a completely new concept of the generation and regeneration of hair, which we presume will contribute a great deal in the field of dermatology. The causative mechanisms of male pattern baldness have been found to be relevant to hereditary factors, hormones, and nutritional factors, and these findings have established a basic concept for the development of completely new hair tonics. It should be noted that the above findings have consequently led to the world-wide approval of various patents related to hair-growing agents.

Postscript

Of all the body odors, axillary odor ranks the highest in the frequency of occurrence; however, there have been no truly effective radical treatment methods until recently. Because the occurrence of bromidrosis is not common among Orientals, development of a radical treatment method has long been hoped for.

More than twenty years have elapsed since we first developed the subcutaneous tissue shaving method for the radical treatment of bromidrosis. Since then we have treated over 30,000 patients and are now very confident that the follow-up results prove to be excellent.

Along with the development of this method, the study on the interaction between the hair follicle and the sebaceous gland has shown remarkable progress. During the course of study on the mechanism of the hair cycle, we have proposed the "Sebaceous Gland Hypothesis" in which it is postulated that the central portion of hair regeneration is sited at the upper isthmal portion of the hair follicle close to the secretory duct opening of the sebaceous gland. It is gratifying to note that the "Essential Hair Cycle" has been newly delineated and differentiated from the "Common Hair Cycle." We have further proposed that the essential hair cycle be divided into four stages—anagen, catagen, telogen, and isthmal—instead of the conventional three. This completely new concept of the hair cycle has led us to develop trichogeous agents which have already proven to be efficacious in accelerating hair growth. A new and innovative study often disrupts traditional concepts. Despite such difficulties, the warm understanding and encouragement shown by such authorities in the field as the late Dr. Hermann Pinkus and Prof. Tai Ho Chung have enabled us to advocate this hypothesis up to the present.

We are most grateful for the support and cooperation of

Springer-Verlag in bringing this new concept into publication and to the attention of all those who are engaged in research studies in this field on an international scale.

We are determined to continue our studies in this field, free from bias, with the hope that the publication of this book will provide a useful contribution to the progress of scientific research.

Acknowledgments. Special thanks are due to Prof. Tai Ho Chung, Assistant Prof. Chung Chul Kim and Yung Chul Choi MD, Biomedical Research Laboratories, Kyungpook Medical School, Taegu, Korea, to Robert M. Nakamura MD, Chairman, Department of Pathology, Scripps Clinic and Research Foundation for their cooperation, and to Mr. William Dyne for the basic preparation of the English manuscript. Also, we would like to acknowledge the cooperation and support of Springer-Verlag, Tokyo, and the assistance rendered by Ms. Yuko Murabayashi, President of MHR Co., Ltd., Ms. Midori Nakagawa, and Ms. Mikiko Azuma, also of MHR Co., Ltd.

References

Abell E, Morgan K (1974) The treatment of idiopathic hyperhidrosis by glycopyrronium bromide and tap water iontophoresis. Br J Dermatol 91:87–91

Abiko J (1986) Feces odor (in Japanese). Nippon Iji-Shinpo 3238:133

Abramson C, Terleckyj B (1979) Bromidrosis: Current concepts related to foot pathology. J Am Podiatry Assoc 69:252–256

Adachi B (1937) Das ohrenschmalz als Rassenmerkmal und der Rassengeruch nebst dem Rassenuntershied der schweissdruseen Z Rassenkunde 6:273–307

Adachi H (1961) Sur le malade qui se croit d'être puant—Etude anthroplogique (in Japanese). Hum Res Juntendo Med J 7:901–917

Adachi K, Kano M (1970) Adenylcyclase in human hair follicle: Its inhibition by dihydrotestosterone. Biochem Biophys Res Commun 41:884–890

Adar R, Kurchin A, Zweig A, Mozes M (1977) Palmar hyperhidrosis and its surgical treatment. Ann Surg 186:34–41

Adson AW, Brown GE (1932) Extreme hyperhidrosis of the hands and the feet treated by sympathetic ganglionectomy. Mayo Clin Proc 7:394

Akobjanoff L, Carruthers C, Senturia BH (1954) The chemistry of cerumen: A preliminary report. J Invest Dermatol 23:43

Amoore J (1977) Specific anosmia and the concept of primary odors. Chem Senses Flavor 2:267

Anderson AS, Fulton JE (1973) Sebum: Analysis by infrared spectroscopy. J Invest Dermatol 60:115–120

Apfelberg DB, Maser MR, Lash H (1976) The use of epidermis over a keloid as an autograft after resection of the keloid. J Dermatol Surg Oncol 2:409–411

Ashby EC, Williams JLO (1976) Cryosurgery for axillary hyperhidrosis. Br Med J 2(6045):1173–1174

Atkins HJB (1949) Peraxillary approach to the stellate and upper thoracic sympathetic ganglion. Lancet II:1152

Baden HP (1987) Disease of the hair and nails. Year Book Medical, Chicago

Baker AB, Baker LH (1975) Clinical neurology, vol 3. Harper and Row, New York

Barber KA, Jackson R (1982) Basic principles of electrolysis.

In: Epstein E, Epstein E Jr (eds) Skin surgery, 5th edn. CC Thomas, Springfield, pp 428–429

Barett EL, Kwan HS (1985) Bacterial reduction of trimethylamine oxide. Annu Rev Microbiol 39:131

Barnstable CJ, Bodmer WF, Brown G, Galfre G, Milstein C, Williams AF (1978) Production of monoclonal antibodies to group A erythrocytes, HLA and other human cell surface antigens. Cell 14:9–20

Bauer WC, Carruthers BSC, Senturia BH (1953) The free amino acid content of cerumen. J Invest Dermatol 21:105–110

Bishop ER (1980) Monosymptomatic hypochondriasis. Psychosomatics 21:731–741

Bodmer JG, Curtoni ES, Van Leeuwen A, Mickey MR, Kjenbye L, Degos L, Botha MC, Wolf E, Mayer WR, Staub-Nilsen L, Piazza A (1975) The ABC and HLA. In: Kissmeger-Nielsen F (ed) Histocompatibility testing. Munksgaard, Copenhagen pp 21–99

Borak J, Eller JJ, Eller WD (1949) Roentgen therapy for hyperhidrosis observation of one hundred and twenty-two patients. Arch Dermatol 59:644

Brand JM, Galask RP (1986) Trimethylamine: The substance mainly responsible for the fishy odor often associated with bacterial vaginosis. Obstet Gynecol 68:682–685

Bretteville-Jansen G (1973) Radical sweat gland ablation for axillary hyperhidrosis. Br J Plast Surg 26:158–162

Chen KC, Forsyt PS, Buchanan TM, Holmes KK (1979) Amine content of vaginal fluid from untreated and treated patients with non-specific vaginitis. J Clin Invest 63:828–35

Choi YC (1988) A study of subcutaneous shaving method as a radical surgical treatment for osmidrosis in 1,200 cases in Korea. J Korea Med Assoc 31(3):299–305

Choi YC, Chung TH, Inaba M (1984a) Determination of tri-acylglyceride in axillary sebaceous gland of osmidrosis patient (in Japanese). Jpn J Aesth Surg (Jap JSAS) 22:150–155

Choi YC, Kim JC, Chung TH, Inaba M, Inaba Y (1984b) A report on the lipid analysis of the axillary dermal surface fat of patients with osmidrosis in Japan and Korea (in Japanese). Jpn J Aesth Surg (Jap JSAS) 23:36–39

Cipollars AC, Crossland PM (1967) X-rays and radium in the treatment of diseases of the skin, 5th edn. Lea and Febiger, Philadelphia

Claycomb CK, Schearer TR (1986) Malodors of the mouth. J Oreg Dent Assoc 55:34–35

Cloward RB (1969) Hyperhidrosis. J Neurosurg 30:544–551

Cohen IK, Beaven MA, Horakova S (1972) Histamine and collagen synthesis in keloid and hypertrophic scar. Surg Forum 23:509

Cohen IK, Bryant CP, Diegelmann RF (1975) Alpha globulin collagenase inhibitors in keloid and hypertrophic scar. Surg Forum 26:61

Cone TE (1968) Diagnosis and treatment, some diseases, syndromes and conditions associated with an unusual odor. Pediatrics 41:993

Crisostomo RF (1989) Hyperhidrosis axillae treated by liposuction curettage. Am J Cosmet Surg 6:117–120

Crissery JT, Rebell GC, Laskas JJ (1952) Studies on causative organism of trichomycosis axillaris. J Invest Dermatol 19:187

Cullen SI (1975) Topical methenamine therapy for hyperhidrosis. Arch Dermatol 111:1158–60

Dausset TJ (1958) Isoleucoanticorps. Acta Haematol 20:156–159

Dohn DF, Sava GM (1978) Sympathectomy for vascular syndromes and hyperhidrosis of the upper extremities. Clin Neurosurg 25:637–650

Dorfman RI, Dorfman AS (1962) Assay of subcutaneous administered androgen on the chick's comb. Acta Endocrinol (Copenh) 41:101–106

Doty RL, Huggins GR (1975) Changes in the intensity and pleasantness of human vaginal odor during the menstrual cycle. Science 190:1316–1318

Doty RL, Kligman A, Leyden J, Orndorth MM (1978) Communication of gender from human axillary odors. Relationship to perceived intensity and hedonicity. Behav Biol 23:373–380

Doty RL, Ram CA, Greene P, Yankell S (1982) Communication of gender from breath odors. Relationship to perceived intensity and pleasantness. Horm Behav 16:13–22

Doty RL, Shaman P, Dann M (1984) Development of the University of Pennsylvania smell identification test. A standardized microencapsulated test of olfactory function. Physiol Behav 32:489–502

Downing DT (1968) Photodensitometry in the thin-layer chromatographic analysis of neutral lipids. J Chromatogr 38:91–99

Downing DT, Strauss JS, Pochi PE (1969) Variability in the chemical composition of human skin surface lipids. J Invest Dermatol 53:322

Downing DT, Stewart ME, Strauss JS (1987) Biology of sebaceous gland. In: Fitzpatrick TB, Eisen AZ, Wolff K, Freedburg IM, Austen KF (eds) Dermatology in general medicine, 3rd edn. McGraw-Hill, New York

Dravnieks A (1975) Evaluation of human body odors: Methods and interpretations. J Soc Cosmet Chem 26:551–557

Durand VJ (1955) Hallucinations olfactives et gustatives. Ann Med Psychol (Paris) 113:777–813

Durward A, Rudall KM (1958) The vascularity and patterns growth of hair follicles. In: Montagna W, Ellis RA (eds) Biology of hair growth. Academic, New York, pp 189–218

Duvic M, Goldsmith LA (1983) HLA and skin disease. In: Goldsmith LA (ed) Biochemistry and physiology of the skin. Oxford University Press, Oxford, pp 951–998

Ellis H (1975) Hyperhidrosis and its surgical management. Postgrad Med 58:191–196

Ellis H (1977) Axillary hyperhidrosis; Failure of subcutaneous curettage. Br Med J 2(6082):301

Ellis RA (1967) Eccrine sweat glands. In: Jadassohn J (ed) Handbuch der Haut und Geschlechtskrankheiten, vol 2. Springer, Berlin Heidelberg New York

Ellis RL, Farr LR, Oberst FW, Crouse CL, Billups NB, Soon WS, Musselman NP, Siddell FR (1974) An apparatus for the detection and quantitation of volatile human effluents. J Chromatogr 100:137–152

Erdos-Brown M (1978) Superfluous hair: Removal with the monopolar diathermy needle. Arch Dermatol Syph 46:496–501

Eriksen SP, Kulkarni AB (1960) Methanol in normal human breath. Science 141:639–640

Frey L (1923) Auriculo-temporal nerve syndrome. Reg Neurol 2:97

Fumiiri M (1968) Surgical treatment of bromidrosis (in Japanese). Jpn J Plast Reconstr Surg 11:61–62

Galtein M (1973) Greffes de peau totale libres. J Chir (Paris) 50:322

Goetz N, Kaba G, Good D (1988) Detection and identification of volatile compounds evolved from human hair and scalp using headspace gas chromatography. J Soc Cosmet Chem 39:1–13

Goodman H (1926) Fox-Fordyce syndrome. Acta Derm Venereol (Stockh) 7:509

Greeley PW (1950) Surgical treatment of chronic suppurative hidradenitis. Arch Surg 61:193–198

Green RS, Downing DT, Pochi PE, Strauss JS (1970) Anatomical variation in the amount and composition of human skin surface lipids. J Invest Dermatol 54:240–247

Greenhalgh RM, Rosengarten DS, Martin P (1971) Role of sympathectomy for hyperhidrosis. Br Med J 1:332

Grice K, Sattar H, Baker H (1972) Treatment of idiopathic hyperhidrosis with iontophoresis of tap water and poldine methosulphate. Br J Dermatol 86:72–77

Guerrero-Santos J (1971) Surgical treatment of axillary hyperhidrosis. Rev San Guadalara 4:49

Haagensen DEJ, Mazoujian G, Dilley WG, Pederson CE, Kister SJ, Wells SAJ (1979) Breast gross cystic disease fluid analysis: Isolation and radioimmunoassay for a major component protein. J Natl Cancer Inst 62:239–247

Hamilton JB (1942) Male hormone stimulation is prerequisite and incitant in common baldness. Am J Anat 71:451–480

Harada J, Inaba M (1985) Histological observation of difference between hircismus-type and nonhircismus-type apocrine gland (in Japanese). Jpn J Aesthetic Surg (Jap JSAS) 23:151–157

Harahap M (1979) Management of hyperhidrosis axillaris. J Dermatol Surg Oncol 5:223–225

Hardaway WA (1879) The treatment of hirsuties. Arch Dermatol 4:337–340

Hartfall WG, Jochimsen PR (1972) Hyperhidrosis of the upper extremity and its treatment. Surg Gynecol Obstet 135:586–588

Hasegawa N (1971) Treatment of postoperative claims made by patients with fixations (in Japanese). Rinsho Fujinka-ka Sanka Clinical Obstetrics and Gynecology 25:121–125

Hasegawa N, Inaba M, Inaba Y (1989) Permanent epilation using light flash (in Japanese). 47th Aesthetic Surgery Congress, Tokyo

Hauser EDW (1945) Diseases of the foot. Saunders, Philadelphia

Hayakawa T (1988) Laws pertaining to esthetic epilation in the world congress to consider medical epilation and esthetic epilation (in Japanese). Jpn Skin-Esthetic Assoc News 8:10

Hayashi N (1958) Distribution and inheritance of the earwax type: Studies on Cerumen. Ochanomizu Igaku Zasshi 6:769–776, 796

Hill BMR (1976) Poldine iontophoresis in the treatment of plantar and palmar hyperhidrosis. Australas J Dermatol 17:92–93

Hinkel AR, Lind RW (1981) Electrolysis, thermolysis and the blend technique: The principles and practice of permanent hair removal. Arroway, Los Angeles, p 47

Hirayama M (1942) Contributions to the study on the nature of cerumen and secretory glands in the external ear canal (in Japanese). Nippon Jibi Inkoka Gakkai Kaiho 48:857–885

Hirota S (1939) Statistic and experimental study in osmidrosis (in Japanese). Nippon Seikeigeka Gakkai Zasshi 58:530–567

Honma H (1925) Über positive Eisenbefunde in den Epithelien der apocrine Schweissdrusen der menschlichen Axillarhaut. Arch Dermatol Syph 148:463–469

Hooft C (1964) The methionine malabsorption syndrome. Lancet II:20

Hurley HJ (1985) Diseases of the apocrine and eccrine sweat glands. In: Moschella SL, Hurley HJ (eds) Dermatology. Saunders, Philadelphia, p 1345

Hurley HJ Jr (1987) Apocrine glands. In: Fitzpatrick TB, Eisen AZ, Wolf KW, Freedberg IM, Austen KF (eds) Dermatol in general medicine, 3rd edn. McGraw-Hill, New York, pp 704–716

Hurley HJ, Shelley WB (1960) The human apocrine sweat gland in health and disease. CC Thomas, Springfield

Hurley HJ, Shelley WB (1963) A simple surgical approach to the management of axillary hyperhidrosis. JAMA 186:109–112

Hurley HJ, Shelley WB (1966) Axillary hyperhidrosis. Clinical features and local surgical management. Br J Dermatol 78:127–140

Imazu K, Yokota I (1931) Injection therapy using formalin. Local injection therapy for osmidrosis: Using formalin solution. Nippon Iji Shinpo 482 2954:19

Ichida K, Ichida T, Ichida H (1949) Studies on the inheritance of earwax types (in Japanese). Jpn J Gent 24:9

Ichihashi T (1936) Effects of drugs on the sweat glands by cataphoresis, and an effective method for suppression of local sweating. J Orient Med 75:101–102

Inaba M (1983) Study on local anesthesia dilution (2nd report). Jpn J Aesthetic Surg (Jap JSAS) 21:122–126

Inaba M (1984) Does lost hair regenerate? Nikkei Science, Tokyo

Inaba M (1985) Can human hair grow again? Baldness: New steps toward prevention and cure. Azabu Shokan, Tokyo

Inaba M (1986) Treatment for hyperhidrosis and bromidrosis (in Japanese), 10th edn. Kinensha, Tokyo

Inaba M (1988) Of what can the hair be formed? (in Japanese). Mainichi Life 1:98–105

Inaba M, Inaba Y (1990a) Male pattern baldness: Sebaceous gland hypothesis. Cosmetics and Toiletries 105:77–87

Inaba Y, Inaba M (1990b) Radical treatment of bromidrosis and hyperhidrosis, especially related to the central portion of the sweat glands (in Japanese). Jpn J Aesth Surg (Jap JSAS) 27:2

Inaba M, Takagi M (1971) Latest report on the operative treatment for bromidrosis: The subcutaneous tissue shaving instrument (in Japanese). Nippon Iji Shinpo 2440:45–47

Inaba M, Matsuyama M, Imaizumi M (1973) Genital bromidrosis (in Japanese). Jpn J Obstet Gynecol 40:69–74

Inaba M, Takagi M, Matsuyama H, Fujinami Y (1974) Radical treatment of osmidrosis axillae. 2nd report: Long-term results of subcutaneous tissue shaving method (Using instrument Type B). Jpn Plast Reconstr Surg 17:300–306

Inaba M, Matsuyama H, Fujinami Y, Takagi C, Nishida N, Shimada N, Takashima M, Mochizuki E (1975) Hereditary nature of osmidrosis (in Japanese). Jpn Med News 2647:48–52

Inaba M, Ezaki T (1977) New instrument for hircismus and hyperhidrosis operation (subcutaneous shaver). Plast Reconstr Surg 59:864–866

Inaba M, Anthony J, Ezaki T (1978a) A new "double tie-over method" and its applications. Aesthetic Plast Surg 2:277–284

Inaba M, Anthony J, Ezaki T (1978b) Preparation of thick tissue sections using cellophane tape, Br J Dermatol 98:625–630

Inaba M, Anthony J, Ezaki T (1978c) Radical operation to stop axillary odor and hyperhidrosis. Plast Reconstr Surg 62:355–360

Inaba M, Ezaki T, Takagi C (1978d) Study on local anesthesia dilution (in Japanese). Cosmetic Medicine (Jap JSAS) (Biyo No Igaku) 6:75–77

Inaba M, Anthony J, McKinstry CT (1979a) Histologic study of the regeneration of axillary hair after removal with subcutaneous tissue shaver. J Invest Dermatol 72:224–231

Inaba M, McKinstry C, Anthony J, Ezaki T (1979b) Epilation by electrocoagulation; factors that result in growth of hair. Jpn J Plast Reconstr Surg 22:29–35

Inaba M, Ojimi T, Abe S, Shimono M (1979c) Experimental results in use of fluor test (Cytocellophane) (in Japanese). Jpn J Obstet Gynecol 46:81–89

Inaba M, McKinstry C, Umezawa F (1981) Regeneration of axillary hair after plucking. J Dermatol Surg Oncol 7:249–259

Inaba M, McKinstry C, Ezaki T (1982) Histologic observation on the increase in density of axillary hair during adolescence. J Dermatol Surg Oncol 8:59–60

Inaba M, Chung TH, Nam JM (1985a) HLA antigens in patients with osmidrosis in Japan. Kyungpook University Med J 26:132–184

Inaba M, Inaba Y, Choi CK (1985b) On the therapy of keloid—especially on the clinical observation of epidermatoplasty. Jpn J Aesthetic Surg (Jap JSAS) 23:87–92

Inaba M, Inaba Y, Chung TH, Kim JC, Choi YC, Kim JH (1987a) Lipid composition of earwax in hircismus. Yonsei Med J 28:49–51

Inaba M, Inaba Y, Choi YC (1987b) The vascular system in the hair follicle (in Japanese). Jpn J Aesthetic Surg (Jap JSAS) 25:94–105

Inaba M, Inaba Y, Choi YC, Tai HC, Song JY (1988a) The study on distribution of estrogen and progesterone receptors in the scalp tissue. Jpn J Aesthetic Surg (Jap JSAS) 25:175–180

Inaba Y, Inaba M, Murakoshi S (1988b) Electrocoagulation in permanent epilation. Proceedings of the 1st international congress of cosmetic surgery, Oslo

Inaba M, Inaba Y, Murakoshi S, Choi YC (1988c) Aesthetic treatment of scar tissue following incomplete surgical procedures for treatment of osmidrosis (in Japanese). Jpn J Aesthetic Surg (Jap JSAS) 25:19–27

Inaba M, Inaba Y, Nakamura K, Choi YC (1988d) Insufficient adhesion following radical surgery for removal of axillary sweat glands. Jpn J Aesthetic Surg (Jap JSAS) 26:10

Ishida A (1958) The remote result of skin resection therapy in osmidrosis axillae (in Japanese). Nippon Rinsho-Hifuka Zasshi 12(4):369–371

Ito M (1936) Osmidrosis in hereditary skin disease (in Japanese). Jitsuken Iho 257:756–760

Ito T (1943) Über den Golgiapparat der ekkrinen Schweissdrusenzellen der menschlichen Haut. Okajimas Folia Anat Jpn 22:273–280

Iwayama S, Fukushima S, Umezawa F, Takeuchi S, Hoshino M, Katsumata H, Nishio H, Suzuki H (1988) Practical anesthesia for aesthetic surgery. Jpn J Aesthetic Surg 25:190–195

Janet P (1903) Les obsessions et la psychasthenic. Felix Alcan, Paris

Jemec B (1975) Abrasio axillae in hyperhidrosis. Scand J Plast Reconstr Surg 9:44

Jensen O, Karlsmark T (1980) Palm-plantar hyperhidrosis: Treatment with alcoholic solution of aluminium chloride hexahydrate. Dermatol 161:133–135

Juhin L, Hansson H (1968) Topical glutaraldehyde for plantar hyperhidrosis. Br J Dermatol 97:327–330

Kanda F, Yagi E, Fukuda M, Nakajima K. Ohta T, Nakata O (1990) Elucidation of chemical compounds responsible for foot malodor. Brit J Dermatol 122:771

Kano K (1952) New treatment of osmidrosis (in Japanese). Operation 6:420, 7:426–430

Kasahara Y, Fuzinawa A, Sekiguchi H, Matsumoto M (1972) Morbid fear of one's own body odor. In: Kasahara Y (ed) Fear of eye-to-eye confrontation and morbid fear of one's own body odor: Mainly in cases of borderline schizophrenia (in Japanese). Igakushoin, Tokyo, pp 33–80

Kataura A, Kataura K (1967) The comparison of free and bound amino acids between dry and wet types of cerumen (in Japanese). Tohoku J Exp Med 91:215–237

Kawabata A (1930) Injection treatment for osmidrosis (in Japanese). Nippon Iji Shinpo, Tokyo

Kawabata I (1964) Electron microscope studies on the human ceruminous gland. Arch Histol Jpn 25:165–187

Kellum RE (1966) Isolation of human sebaceous glands. Arch Dermatol 93:610–612

Kellum RE (1967) Human sebaceous gland lipids. Analysis by thin-layer chromatography. Arch Dermatol 59:218–220

Kimura T, Hagiwara Y (1985) Regulation of urine marking in male and female mice: Effects of sex steroids. Horm Behav 19:64–70

Kimura T, Daumae M (1988) Factors regulating urination patterns in male and female mice (*Mus Musculus*). Zoological Science 5:855–861

Klein JA (1986) Absorption pharmacokinetics of the tumescent technique for local anesthesia and vasoconstriction of subcutaneous fat in liposuct surgery. Proceedings of the 2nd world congress of liposuction surgery. Philadelphia

Kligman AM, Shehadeh AN (1964) Pubic apocrine glands and odor. Arch Dermatol 89:461–463

Knudsen EA, Neier CNK (1963) Treatment of hyperhidrosis with topical propantheline bromide. Acta Derm Venereol (Stockh) 43:154–157

Kobayashi T (1984) Electrosurgery using insulated needles (part 1). The 27th congress of the Jpn Soc of Plast and Reconstr Surg, Tokyo

Kobayashi T (1985) Electroepilation using insulated needles: Epilation. J Dermatol Surg Oncol 11:993–1000

Kobayashi T (1988) Electrosurgery using insulated needles: Treatment of axillar bromhidrosis and hyperhidrosis. J Dermatol Surg Oncol 14:749–752

Kobori T, Narumi J (1958) Clinical dermatology (in Japanese). Treatment of osmidrosis axillae with new mercurial compound. Rinsho Hifu Hinyoki (Clin Dermatol Urol) 12:361–368

Komatsu T, Tajima S, Nishikawa T (1989) An immunohisto-chemical study of normal sweat glands using an antibody to the breast cyst fluid protein (GCDEP-15). Jpn Dermatol 99: 885–989

Koyama M (1925) Tokyo medical newsletter. Tokyo Medical Journal 2451:2722–2723

Kuno Y (1956) Human perspiration. CC Thomas, Springfield

Kuroda K (1957) Follow-up results for the treatment of osmidrosis (in Japanese). Hifuka Seibyoka Zasshi 64(6):421

Kurosumi K (1977) Fine structure of the human sweat ducts of eccrine and apocrine type. Arch Histol Jpn 40:203–234

Kurosumi K, Kitamura T, Iijima T (1959) Electron microscope studies on the human axillary apocrine sweat glands. Arch Histol Jpn 523

Kurosumi K, Kurosumi U, Tosaka H (1982) Ultrastructure of human sweat glands with special reference to the transitional portion. Arch Histol Jpn 45:213–238

Kurosumi K, Shibasaki S, Ito T (1984) Cytology of the secretion in mammalian sweat glands. Int Rev Cytol 87:253–329

Kushima K (1966) Psychological study of pregnant patients (in Japanese). Seishinsho J 6:156–161

Labows JN (1979) Human odors—what can they tell us? Perfumer Flavorist 4:12–17

Labows JN (1988) Odor detection, generation, and etiology in the axilla. In: Felger C, Laden K (eds) Antiperspirants and deodorants, Marcell Dekker, New York pp 321–343

Labows JN, McGinley KJ, Leyden JJ, Webster GF (1979a) Characteristic r-lactone odor production of the genus *Pityrosporum*. Appl Environ Microbiol 38:412–415

Labows JN, Preti G, Hoelze E, Leyden J, Kligman A (1979b) Axillary odors: Compounds of exogenous origin. J Chromatogr 163:294–299

Labows JN, McGinley KJ, Kligman AM (1982) Perspectives on axillary odor. J Soc Cosmet Chem 34:193–202

Labows JN (1991) Personal communication

Landes VE, Kappesser HJ (1979) Zur operativen Behandlung der hyperhidrosis axillaris. Fortschr Med 97:2169

Lee CW, Yu JS, Turner BB, Murray KE (1976) Trimethylaminuria: Fishy odor in children. N Engl J Med 295:937

Lerner C (1942) Treatment of hypertrichosis by electrocoagulation NY State J Med 42:879–882

Levine IM, Harris OJ (1955) Chemical sympathectomy: New approach to treatment of localized hyperhidrosis. Arch Dermatol Syph 71:226–230

Levit F (1980) Treatment of hyperhidrosis by tap water iontophoresis. Cutis 26:192–194

Levy DS, Salter MM, Roth RE (1976) Postoperative irradiation in the prevention of keloids. AJR 127:509

Leyden JJ, Shalita AR (1986) Rational therapy for acne vulgaris: An update on topical treatment. J Am Acad Dermatol 15: 907–914

Leyden JJ, McGinley KJ, Holzle E, Labows JN, Kligman AM

(1981) The microbiology of the human axilla and its relationship to axillary odor. J Invest Dermatol 77:413–416

Lu DP (1982) Halitosis: An etiologic classification, a treatment approach and prevention. Oral Surg 54:521–526

Lucky AW (1988) The paradox of androgens and balding: Where are we now? J Invest Dermatol 91:99–100

Mandour MA, El-Ghazzawi EF, Toppozada HH, Malaty HA (1974) Histological and histochemical study of the activity of ceruminous glands in normal and excessive wax accumulation. J Laryngol Otol 88:1075–1085

Marples RR, Downing DD, Kligman AM (1971) Control of free fatty acids in human surface lipids by *corynebacterium acnes*. J Invest Dermatol 56:127–131

Marton MH (1940) Treatment of hypertrichosis by improved apparatus and technique. Arch Phys Therapy 21:678–683

Massler M, Ensline RD, Bolden TE (1951) Fetor exore Oral Surg 4:110–125

Matsunaga E (1962) The dimorphism in human normal cerumen. Ann Hum Genet 25:273–286

Matsunaga E, Itoh S, Suzuki T, Sugimoto R (1954) Incidence and inheritance of ear-wax types (in Japanese). Sapporo Med J 6:368

McKinstry CT, Inaba M (1979) Epilation by electrocoagulation; factors that result in regrowth of hair. J Dermatol Surg Oncol 5:407–411

Meech RJ, Loutit J (1985) Non-specific vaginitis; diagnostic features and responses to imidazole therapy. NZ Med J 98:389–91

Michael RP, Bonsall RW, Warner P (1974) Human vaginal secretions: Volatile fatty acid content. Science 186:1217–1219

Michel CE (1875) Trichiasis and districhiasis with an improved method for their radical treatment. St. Louis Clinical Record 2:145–148

Milson JP, Craig RDP (1973) Collagen degradation in cultured keloid and hypertrophic scar tissue. Br J Dermatol 89:635

Mitsumine K, Okugawa S (1940) Histological research for bromidrosis in the pubic area (in Japanese). Hifu Hinyo Shi 48(4):365

Miyake S (1932) The hereditary factor in wet earwax (in Japanese). Keijo Isen Kyo 13:192

Miyamoto Y (1976) Self-consciousness (olfactory paranoia): Symptomatic consideration (in Japanese). Rinsho Seishingaku (Clin Psychol Med) 5:1223–1230

Montagna W, Ellis RA (1957) Histology and cytochemistry of human skin. JNCI 69:451–463

Montagna W, Parakkal PF (1974) The structure and function of skin, 3rd edn. Academic, New York

Montagna W, Chase HB, Lobitz WC Jr (1953) Histology and cytochemistry of human skin. J Invest Dermatol 20:415–423

Moretti G (1965) Das Haar. In: Stüttgen G (ed) Die normale und pathologishe Physiologie der Haut. Gustar Fisher, Stuttgart, pp 506–553

Morita G (1947) On the inheritance of wet cerumen among Japanese (in Japanese). Juzenkai Zasshi 50:54–60

Moriyama G (1927) Microchemical study in axillary sweat glands especially in osmidrosis patients (in Japanese). Nagasaki Igakukai Zasshi 5:302–310

Munger BL (1971) Histology and cytology of the sweat glands. In:

Helwig EB, Mostofi FK (eds) The skin. Williams and Wilkins, Baltimore, pp 47–64

Murata M (1960) Clinical observation in osmidrosis (first report). Jpn J Dermatol 70:969

Nag W (1984) Reaction of H_2S and CH_3SH with oral mucosae. J Dent Res 63:263

Nagai T (1935) On the relationships between axillary odor, earwax types, and blood groups (in Japanese). Kyoto Med J 32:511

Nagamitsu G (1941) Pathologisch-histologische Untersuchungen uber die osmidrosis axillae (in Japanese). Keio Igaku 21: 1011–1039

Nagashima T (1934) Ceruminous glands and cerumen in Japanese, with special reference to axillary odor. Jpn J Dermatol 36:690

Nakahira K (1957) Follow-up results of curettage method for treatment of osmidrosis. Jpn J Dermatol 67(6):421

Nakajima A, Hirano I (1968) Distribution and inheritance of earwax types. Jpn J Human Genet 13:201–207

Nieminen E, Leikola E, Koljonen M, Kiistala U, Mustakallio KK (1967) Quantitative analysis of epidermal lipids by TLC with special reference to seasonal and age variation. Acta Derm Venereol (Stockh) 47:327–338

Nikkari T, Valavaara M (1970) The influence of age, sex, hypophysectomy and various hormones on the composition of the skin surface lipid of the rat. Br J Dermatol 84:459–472

Nishimura M, Miyasaka M, Yano R, Yamaguti H, Morita Y, Yamada H, Saito H, Kayama H, Nagata M (1986) More radical operation for bromidrosis (in Japanese). Jpn J Plast Reconstr Surg 6:707

Nitta H, Ikai K (1954) Studies on body odor, part 1. Separation of the lower fatty acids of cutaneous excretion by paper chromatography. Nagoya J Med Sci 1:217–224

Nozoe M (1943) Statistical observations on wet cerumen (in Japanese). Keijo Isen Kiyo 13:189

Ono I, Ohura T, Hamamoto J, Sugihara T, Umeda T (1978) Fibrinolytic therapy on hematoma beneath grafted skin. Jpn J Plast Reconstr Surg 21:563–568

Ozaki T, Hiruta K, Kameda N (1953) Study of new surgical treatment for osmidrosis. Operation VII(5):430–434

Padgett EC (1939) Calibrated intermediate skin grafts. Surg Gynecol Obstet 69:799

Payne R, Rolfs MR (1958) Fetomaternal leukocyte incompatibility J Clin Invest 37:1756–1763

Petrakis NL, Wiesenfeld SL, Flander L (1986) Possible influence of age on the expression of the heterozygous cerumen phenotype. Am J Phys Anthropol 69:437–440

Pheifer TA, Forsyth PS (1978) Nonspecific vaginitis: Role of *Haemophilus vaginalis* and treatment with metronidazole. N Engl J Med 298:1429–1434

Pinkus H (1969) Sebaceous cysts are trichilemmal cysts. Arch Dermatol 99:544

Pinkus H, Mehregan AH (eds) (1981) The guide to dermato-histopathology, 3rd edn. Appleton-Century-Crofts, East Norwalk, pp 32–33

Preti G, Huggins GR (1975) Cyclical changes in volatile acidic metabolites of human vaginal secretions and their relation to ovulation. J Chem Ecol 1:361–376

Price GD, Smith N, Carlson DA (1979) The attraction of female mosquitos to stored human emanations in conjunction with adjusted levels of relative humidity, temperature and carbon dioxide. J Chem Ecol 5(3):383–395

Ramasastry P, Downing DT, Pochi PE (1970) Chemical composition of human skin surface lipids from birth to puberty. J Invest Dermatol 54:139

Reese JD (1949) Dermatape: A new method for the management of split skin grafts. Plast Reconstr Surg 1:98–105

Reisner RM, Puhvel M (1969) Lipolytic activity of *Staphylococcus albus*. J Invest Dermatol 53:1–7

Richter VJ, Tonzetich J (1964) The application of instrumental technique for the exudation of odoriferous volatile from saliva and breath. Arch Oral Biol 9:47–53

Ridley CM (1969) A critical evaluation of the procedures for treatment of hirsutism. Br J Dermatol 81:146–153

Rigg BM (1977) Axillary hyperhidrosis. Plast Reconstr Surg 59:334–342

Rizzo AA (1967) The possible role of hydrogen sulfide in human periodontal disease. Pedodontics 5:233–236

Rostenberg A (1925) Epilation with diathermy. Med J Rec 121:751

Rothman S (1954) Physiology and biochemistry of the skin. University of Chicago Press, Chicago

Roussel Laboratories (1958) Sofratulle (gauze including fradiomycin sulfate). Therapie 13:945

Russell JD, Witt W (1976) Cell size and growth characteristics of cultured fibroblasts in isolated normal and keloid tissue. Plast Reconstr Surg 57:207

Sastry S, Buck K, Janak J, Dressler M, Preti G (1980) Volatiles emitted by humans. In: Waller G, Dermer OC (eds) Biochemical applications of mass spectrometry. Wiley, Chichester, pp 1086–1133

Sato K, Dobson RL, Mall JWH (1971) Enzymatic basis for the active transport of sodium in eccrine sweat gland: Localization and characterization of Na-K-ATPase. J Invest Dermatol 57:10–16

Sato Y (1976) The hair cycle and its control mechanism: In: Montagna W, Kobori T (eds) Biology and Disease of the Hair University of Tokyo Press, Tokyo, pp 3–13

Savill A, Warren C (1962) The hair and scalp, 5th edn. Arnold, London p 304

Sawaya ME, Hong LS, Garland LD, Hsia SL (1988) Hydroxysteroid dehydrogenase activity in sebaceous glands of scalp in MPB. J Invest Dermatol 91:101–105

Sawaya ME, Hong LS, Hsia SL (1989) Increased androgen binding capacity in sebaceous glands in scalp of male pattern baldness. J Invest Dermatol 92:91–95

Schiefferdecker P (1917) Die Hautdrüsen des Menschen und des Saügetieres, ihre biologische und rassenanatomische Bedeutung, sowie die muscularis sexualis. Biol Zentr 37:534–562

Schiefferdecker P (1922) Die Hautdrüsen des Menschen und des Saügetieres, ihre Bedeutung, sowie die muscularis sexualis. Zoologica 72:1–154

Scheimann LG, Knox G, Sher D, Rothman S (1960) The role of bacteria in the formation of free fatty acids on the human skin surface. J Invest Dermatol 34:171–174

Schwartz IL, Thaysen JH (1956) Excretion of sodium and

potassium in human sweat. J Clin Invest 35:114–120

Schweikert HV, Wilson JD (1974) Regulation of human hair growth by steroid hormones: Testosterone metabolism in isolated hair. J Clin Endocrinol Metab 38:811–819

Scriver CR, Rosenberg LE (1973) Amino acid metabolism and its disorders. Saunders, Philadelphia

Serikawa K, Tokuhashi I, Murakami M, Ikari Y, Kanda S, Shimoda N (1988) A case of eccrine spiradenoma. Jpn J Dermatol 98:879–889

Shehadeh N, Kligman AM (1963) The effect of topical antibacterial agents on the bacterial flora of the axilla. J Invest Dermatol 40:61–67

Shelley WB (1951) Apocrine sweat. J Invest Dermatol 17:255

Shelley WB (1954) Experimental miliaria in man: The effect of poral closure on the secretory function of eccrine sweat glands. J Invest Dermatol 22:267–271

Shelley WB, Florence R (1960) Compensatory hyperperidrosis after sympathectomy. N Engl J Med 263:1056

Shelley WB, Horvath PN (1951) Comparative study of the effect of anticholinergic compounds on sweating. J Invest Dermatol 16:267–274

Shelley WB, Hurley HJ, Nicholas AC (1953) Axillary odor: Experimental study of the role of bacteria, apocrine sweat and deodorants. Arch Dermatol 68:430

Shenag SM, Psira M (1987) Treatment of bilateral axillary hyperhidrosis by suction-assisted lipolysis technique. Ann Plast Surg 19:548–551

Shichijo S, Masuda H, Takeuchi M (1979) Carbohydrate composition of glycopeptides from human cerumen. Biochem Med 22:256–263

Shih CJ, Wang YC (1978) Thoracic sympathectomy for palmar hyperhidrosis: Report of 457 cases. Surg Neurol 10:291–296

Shikano T (1960) Clinical study of chronic hallucinative disorder (in Japanese). Seishinigaku (Psychol Med) 2:37–41

Shimada N (1953) Trial standardization of neurosis research. Jpn J Appl Psychol Soc 1:63

Shimada N, Takashima M, Inaba M, Mochizuki E (1974) The psychosomatic approach for osmidrosis patients (in Japanese). Nippon Iji Shinpo 2619:30–34

Shirakabe T, Kawata M, Kinugasa T, Shirakabe I (1986) Easy osmidrosis treatment using a special cannule for liposuction. Jpn J Plast Reconstr Surg 6:707–708

Sho Z (1969) Pedicle flap surgery for treatment of osmidrosis. Plast Esthetic Surg 3(4):277

Shrivastava SN, Singh G (1977) Tap water iontophoresis in palmaoplantal hyperhidrosis. Br J Dermatol 96:189–195

Siakotos AN, Goebel HH, Patel V, Watanabe I, Zeman W (1972) The morphogenesis and biochemical characteristics of ceroid isolated from cases of neuronal ceroid-lipofuscinosis. Adv Exp Med Biol 19:53–61

Sidbury JB Jr, Smith EK, Harlan W (1967) An inborn error of short chain fatty acid metabolism. The odor of sweaty feet syndrome. J Pediatr 70:8

Skoog T, Thyreson N (1962) Hyperhidrosis of the axillae: A method of surgical treatment. Acta Chir Scand 124:531–538

Sloan JB, Soltani K (1986) Iontophoresis in dermatology. J Am Acad Dermatol 15:671–684

Solomons B (1962) Congenital hyperhidrosis of palms and soles. Excerpta Med Int Cong Ser 52:98–99

Speigel CA, Amsel R, Eschenbach D, Schoenknect F (1980) Anaerobic bacteria in non-specific vaginitis. N Engl J Med 303:601–606

Stewart ME (1982) Variability in the fatty acid composition of wax esters from vernix caseosa and its possible relation to sebaceous glands. J Invest Dermatol 78:291

Stillians AW (1916) The control of localized hyperhidrosis. JAMA LXVII (27):2015

Stolman LP (1987) The treatment of excess sweating of the palms by iontophoresis. Arch Dermatol 123:893–896

Strauss JS, Kligman AM (1956) The bacteria responsible for apocrine odor. J Invest Dermatol 27:67

Sulser GE, Brening KH, Fosdick LS (1939) Some conditions that affect the odor concentration of the breath. J Dent Res 18: 355–359

Suzuki A (1960) Heredity of the larygol-otological area (in Japanese). Iden Igaku 29:637

Suzuki M, Suzuki A, Yamakawa T, Matsunaga E (1985) Characterization of 2.7-anhydro-N-acetylneuraminic acid in human wet cerumen. J Biochem 97:509–515

Suzuki Y (1938) Concerning wet ear wax and osmidrosis (in Japanese). Nippon Jibi-inkoka Gakkai Kaiho 11:1087–1092

Swerdlow DB, Salvati EP, Rubin RJ, Labow SB (1974) Electrosurgery: Principle and use. Dis Colon Rectum 17(4):482–486

Tagg JR, Dajani AS, Wannamaker LW (1976) Bacteriocins of gram-positive bacteria. Bacteriol Rev 40:722

Takagi S (1974) The story of the sense of smell (in Japanese). Iwanami Shoten, Tokyo

Takagi S (1988) Sense of smell and function of odors (in Japanese). Fragrance J 91:15–18

Takami Y (1960) Supplementary study of bromidrosis (in Japanese). Jpn J Dermatol 70:1266–1293

Takamiya A (1952) Statistical observation on wet cerumen (in Japanese). Seibutsu Tokei Shi 1:1–4

Takatsu GD (1957) Surgical treatment of hyperhidrosis. AMA Arch Dermatol 76:31–38

Takayasu S, Wakimoto H, Itami S, Sano S (1980) Activity of testosterone 5α-reductase in various tissues of human skin. J Invest Dermatol 74:187–191

Tamura H (1915) Die histologische Untersuchung der osmidrosis. Jpn J Dermatol (Hifu-Hinyo Shi) 15:388

Tanioku K (1956) Operation method for osmidrosis, 2nd edn (in Japanese). Kanehara Shuppan, Tokyo

Telford ED (1938) Sympathetic denervation of the upper extremity. Lancet 1:70

Terasaki PI, McCelland JD (1964) Microdroplet assay of human serum cytotoxins. Nature 204:998–1000

Thurnon F, Ottenstein B (1952) Studies on the chemistry of human perspiration with especial reference to its lactic acid content. J Invest Dermatol 18:333–339

Tonzetich J (1971) Direct gas chromatography in mouth air in man. Arch Oral Biol 16:587–597

Tonzetich J (1978) Oral malodour: An indication of health status and oral cleanliness. Int Dent J 28:308–319

Tonzetich J, Richter VJ (1964) Evaluation of volatile odoriferous compounds of saliva. Arch Oral Biol 9:39–45

Tsuji K, Komori K, Yasuda N (1979) Proceedings of the first Asia and Oceania histocompatibility workshop and conference. Japan HLA Association, Tokyo, p 113

Tsunoda M (1975) Gastrochromatographic analysis of halitosis in expiration (in Japanese). Nishi Shushi (Jpn Dental Weekly) 17:1–13

Tsunoda M, Watanabe Y (1988) Development of a breath detector for halitosis (in Japanese). Nishi Shushi (Jpn Dental Weekly) 30(4):1127–1140

Umezu M (1975) Gas chromatography analysis of the exhalation of halitosis patients (in Japanese). Nishi Shushi. (Jpn Dental Weekly) 17:1–13

Van Rood JJ, Eernisse JG, Van Leeuwen A (1958) Leukocyte antibodies in sera from pregnant females. Nature 8:1735–1736

Wagner RF, Tomich JM, Grand DJ (1985) Electrolysis and thermolysis for permanent hair removal. J Am Acad Dermatol 12:441–449

Walter K (1965) Über das "Phobische Beziehungssyndrom." Nervenarzt 36:7–11

Watanabe K, Akagawa T, Arai K (1967) Pressure fixation methods for skin grafts. Jpn J Plast Reconstr Surg 10(4):305–312

Weber TB, Rhoades JW (1960) Detection of trace constituents in oral fluid by gas chromatography. J Derm Res 40:750

Welch E, Geary J (1984) Current status of thoracic dorsal sympathectomy. J Vasc Surg 1:202–214

Wilkinson HA (1984) Radiofrequency percutaneous upper thoracic sympathectomy. N Engl J Med 311:34–36

Williams LL, Howard RS (1983) Lipofuscin in rabbit skin. Arch Pathol Lab Med 107:40–45

Winner's Japan Co., Ltd. (1988) Digital automatic mouth odor detector (DE 160). Operation Manual, p 1

Yamada K (1932) Quantitative Untersuchung der Anhangsorgane der Haut bei dem Deutschen. Folia Ant Jap 10:721

Yamamura Y (1981) Body odor. In: Yamura Y, Kukita J, Sano E (eds) Handbook in dermatology, 2nd edn. Nakayama Shoten, Tokyo, p 163

Yano K, Hata Y, Hosokawa K, Matsuka K (1988) Treatment of osmidrosis by the Skoog method. Jpn J Plast Reconstr Surg 31(10):908–914

Yoshida J (1963) Radical surgical treatment for osmidrosis using drum dermatome (in Japanese). Jpn J Dermatol 73:384–485

Yoshida Z (ed) (1987) The story concerning sweating and odor injection therapy for osmidrosis and hyperhidrosis. Wako Insatsu, Tokyo

Yoshihiro I (1942) Clinical and pathologic findings on bromidrosis (in Japanese). Jpn J Dermatol 51:538–556

Youmans GP, Paterson PY, Sommers HM (1975) The biological and clinical basis of infectious diseases, 1st edn. Saunders, Philadelphia

Zeng X, Leyden J, Lawley H, Sawano K, Nohara I, Preti G (1991) Analysis of the characteristic odors from the male axillae. J Chem Ecol 17:1469–1492

Index